The Sense of Smell
and its Abnormalities

To Nicole

The Sense of Smell and its Abnormalities

Ellis Douek, F.R.C.S.

CONSULTANT OTOLOGIST GUY'S HOSPITAL, LONDON

CHURCHILL LIVINGSTONE
Edinburgh & London 1974

CHURCHILL LIVINGSTONE
Medical Division of Longman Group Limited

Distributed in the United States of America by
Longman Inc., New York and by associated
companies, branches and representatives throughout
the world.

First published 1974

ISBN 0 443 00950 3
Library of Congress Catalog Card Number 73-84621

Printed in Great Britain

Preface

The study of the chemical senses has been stimulated considerably over the past decade by the interest shown in agriculture and in the perfume and flavour industries. The doctor who has to deal with abnormal responses, and who could therefore make a contribution to this field, has tended to remain isolated. Indeed, workers in different disciplines have suffered from a marked degree of separation until the recent formation of the European Chemo-Reception Organisation (ECRO), and this book is intended to be a contribution to the process of better mutual information.

The author is only too aware of the deficiencies which an attempt to bring different fields together in a book such as this must uncover. However the need to have even a superficial knowledge of the whole of olfaction in its broadest sense is now so urgent that a step such as this seems justified. It is to be hoped that each specialist will accept any weakness in his own field in return for information on others.

Many people have given help and advice and I have tried to give credit in the text. I owe much to John Ballantyne in so many ways, that to record my debt to him for suggesting this book and making it possible, can surprise no one who knows him.

My friend Mary Lewis helped me with the Chapter on Smell and Culture and I acknowledge this with pleasure.

My Senior Registrar, Mr R. Parker, and my Research Assistant, Mr W. P. R. Gibson read the proofs, and I am most grateful to Nancy Hibbs who so patiently prepared the manuscript.

I need not add that the information collected here comes from a mixture of personal experience, communications from colleagues, books and journals of such a varied nature that they are unlikely to be seen in a cross-disciplinary manner.

1974   ELLIS DOUEK

Contents

CHAPTER ONE

The Structure and Function of the Nose

STRUCTURE

If the bodily structures of man are compared with those of other animals it is their extraordinary similarity which is at first more striking. Later the differences between the orders and species—differences arising from development and specialization—begin to place themselves in an evolutionary pattern. In animals such as the dog, for instance, the skeleton of the snout is very extensive, projecting forwards well beyond the orbits. It consists of the maxilla and premaxilla into which are embedded the upper teeth, and contains the bony skeleton of the nose which is lined by a delicate membrane part of which forms the organ of smell. Although the elements of the skull are still present, the relative proportions and shape alter considerably with the evolution of the highest animal Order—the Primates. These changes were associated with the move to arboreal life, and apart from man and some heavy gorillas it is in the trees that the Primates have remained. Dr G. G. Simpson has divided this Order into two sub-Orders: Prosimii which includes the tarsiers, lemurs and tree-shrews, while the others comprising man, anthropoid apes and monkeys form the Anthropoidea because of their man-like appearance. Some of the outstanding features which produce this resemblance involve the skeleton of the face, which is more flattened than that of the lower mammals such as the dog with a larger, somewhat bulging cranium and close-set, forward looking eyes.

Arboreal existence created a totally different environment with which anatomical changes were associated. The appearance of a real hand as opposed to a fore-foot and its increasing use in grasping was accompanied by a dwindling of the importance of the teeth and therefore a reduction of the strength and size of the snout. Olfaction became less important for life in the trees than on the multi-odorous surface of the earth. This was particularly the case as vision had been considerably elaborated by the increased range

and horizon produced by elevation and the appearance of stereo-
scopic vision.

Thus the sense of smell, although still fairly well developed in
the Prosimii gradually lost its importance in the apes and man and
this was reflected by anatomical changes. In the dog the maxillo-
turbinal is greatly increased in surface area by complex branching and the olfactory area is very extensive. In the higher apes and
man this area is very small and the ectoturbinals, or outer row of
ethmo-turbinals, is represented only by the ethmoidal air cells. Thus
the general reduction in this nasal apparatus contributed to the
reduction in prominence of the whole region.

The progressive increase in the size of the brain brought about
the enlargement of the part of the skull encasing it relative to the

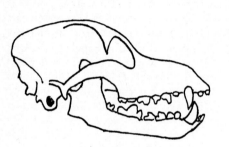
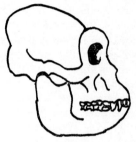

Dog   Simian

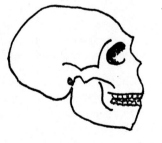
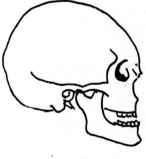

Neanderthal   Man

Figure 1.1   Outlines of skulls.

skeleton of the face. This disproportionate development of the cranium produced a change in its position with regard to the face, so that in man the latter appears to have withdrawn below it (Fig. 1.1).

The nose itself is associated with both olfaction and respiration, so that its structure and function is an expression of their close relationship, and the various factors which affect the sense of smell cannot be considered without an understanding of the main features of nasal anatomy and physiology.

The cavity of the nose lies between the nostrils and the naso-pharynx. This is a space behind the nose which leads downwards to the pharynx and larynx. The nasal cavity is divided into two halves by a partition, the nasal septum, and the two openings which form the back of the cavity are known as the posterior choanae. Their size relative to the nostrils is an important factor in the formation of air currents through the nose, and will be discussed with these problems.

The part of the nose which projects from the face is known as the external nose.

The external nose

The skeleton of this structure is mainly cartilaginous, particularly in its anterior part, and this gives it considerable mobility. Two small nasal bones form the bony part of this framework. They lie one on either side of the septum, articulating with the maxillary and frontal bones and forming the highest part or bridge of the external nose. The cartilaginous skeleton consists of two plates of hyaline cartilage on each side, the upper one articulating with the nasal bones and the cartilage of the septum between them. The opening of the nostril itself is bounded on either side by a small alar cartilage which contains a large proportion of elastic fibres making it more easily compressed or dilated, thus altering the size of the opening. This opening is also known as the anterior nares, or less commonly, the anterior choanae. The two halves are separated by the collumela which forms the front part of septal cartilage.

The skin of the external nose continues to line the inside of the nostrils for about 2 cm before it joins the mucosa in that part of the nose called the vestibule.

The alae of the nose may be moved by a number of muscles: the compressor and the *dilator naris*, and the *levators anguli oris* and *labii superioris alaeque nasi*. Occasionally these muscles are inadequately developed and this produces narrow, slit-like nostrils with weak, thin alae which can be sucked in by the fall in intranasal pressure during inspiration.

The attachment of the external nose to the skull forms an opening which is pyriform in shape. Its boundaries are the tooth-bearing part of the maxilla below and on either side the stout nasal processes of the maxillae which articulate with the frontal bones. In the angle formed by these joints lie the nasal bones.

The external nose receives its blood supply mainly from the facial and maxillary arteries and it drains into the anterior facial vein which communicates with the ophthalmic veins and thus with the intracranial venous sinuses.

The nerve supply is from the infratrochlear nerve above, the external nasal nerve, and branches of the infraorbital nerve around the nostrils. All these are terminal branches of the sensory part of the fifth (trigeminal) cranial nerve.

The shape of the external nose is one of the most obvious racial characters which have divided mankind and it is not surprising that anthropologists have attempted to classify these shapes in a statistical manner. One index thus calculated is the:

$$\frac{\text{Width of the pyriform opening} \times 100}{\text{Height}}$$

This height is measured from the nasion (junction between nasal and frontal bones) and the floor of the pyriform opening. The value of the index is limited to the study of skulls and does not apply to the external nose proper. A second nasal index is the:

$$\frac{\text{Greatest breadth} \times 100}{\text{Height}}$$

This index has been used to classify shapes of nose:

Leptorrhine	—	White races	Below 70
Mesorrhine	—	Asiatic origin	70 to 85
Platyrrhine	—	Black races	Over 85

All attempts made to correlate the shape of the nose with olfaction have been unsuccessful and no evidence has been presented showing that the sense of smell has a racial variation. On the other hand there appears to be some association with climatic differences. According to Thomson and Dudley Buxton (1923), the best correlation is with relative humidity; thus the long narrow nose which breaks up the airflow is more suitable in a cold, dry climate while a nose which allows a free stream would be acceptable in a hot, moist atmosphere. Others have disputed this and sugest that absolute humidity unrelated to temperature gives us close a correlation (Wiener, 1954).

The nasal cavity

The floor of the nose is formed by the hard palate: a plate of maxilla in front and the horizontal plate of the palatine bone behind (Fig. 1.2).

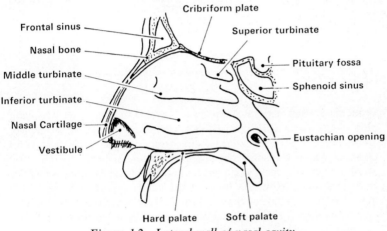

Figure 1.2 Lateral wall of nasal cavity.

The roof is in three portions: a middle horizontal portion formed by the cribiform plate of the ethmoid, and two sloping portions. The anterior portion slopes downwards beneath the frontal sinus and the bridge of the nose which is formed by the nasal bones; the posterior portion slopes backwards beneath the sphenoid sinus (Fig. 1.3).

The cribriform plate is a perforated, flat bony bridge between the ethmoidal labyrinths which lie in each lateral wall of the nose and contain the cells of the ethmoidal sinuses. It is ridged in the midline and this ridge is continued anteriorly by an upwards triangular projection which resembles a cock's comb—the crista galli. To this is attached the falx cerebri which is the sickle-shaped process of

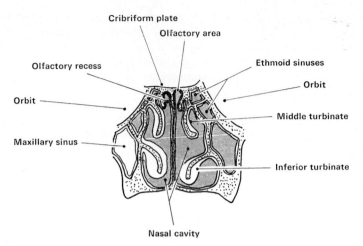

Figure 1.3   Coronal section through face.

dura mater that separates the two cerebral hemispheres. On each side of the crista galli is a slit-like fissure into which slips another dural process. A little lateral to the front part of this fissure is an opening which transmits the anterior ethmoidal nerve to the nasal cavity. The cribriform plate on both sides of the midline is narrow and depressed below the floor of the anterior cranial fossa. On it rests the gyrus rectus of the cerebrum and it is perforated by numerous foramina for the olfactory nerves. These foramina decrease in number and diameter with increasing age, and it has been suggested that this is the cause of presbyosmia, a gradual failure of the sense of smell which occurs in old age. This mechanistic view of bony encroachment overlooks the possibility that it is a primary atrophy of the nerves which leads to a closure of their ducts.

The lateral walls of the nasal cavity seperate it from the maxillary sinus below and the series of intercommunicating cells which form the ethmoidal sinus above. These air-containing sinuses are lined by the same mucosa that lines the nasal cavity. They are not totally enclosed as they communicate with the nose through small openings in its wall which are protected by bony projections called turbinates or conchae. There are three of these turbinates, a superior, a middle and an inferior one. The superior one is small and high up on the lateral wall, and as it projects backwards as well as outwards it appears to merge in front with the anterior part of the middle turbinate when these are covered by soft tissue. The inferior turbinate is the largest and contains, beneath the mucosa, large vascular spaces fed by small arteries and emptied into a venous plexus. This system is under autonomic nervous control and can rapidly swell as it becomes engorged with blood, thus forming a highly sensitive erectile tissue.

The nasal septum is formed by cartilage in front and bone behind. The two bones involved are the vertical plate of the ethmoid, and forming the septum between the posterior choanae, the vomer. This last bone leads backwards and upwards to the sphenoid sinus. This septum is rarely entirely straight and severe deflections are relatively common, obstructing one or both nasal cavities.

Immediately above the incisive canal at the lower edge of the cartilaginous septum can be found a small pit. Sometimes this contains a minute opening leading backwards into a small blind pouch. This is a vestigial structure and appears to have no function in man, but it represents the vomero-nasal organ of Jacobson which in lower animals is supplied by branches of the olfactory nerve and has a role in the sense of smell.

Histology

The small area just inside the nostrils is known as the vestibule and is lined by skin. This skin bears large strong hairs, the vibrissae, which may have a filtering and protective role.

The rest of the nasal cavity is lined by mucous membrane. With the exception of the olfactory area which will be described in the following chapter, the mucosa is almost entirely columnar ciliated epithelium. This type of so-called 'respiratory epithelium' lines, with interruption, the parnasal sinuses and the nasopharynx to the level of the upper border of the superior constrictor muscle, coating the region of the Eustachian tubes. The extreme anterior region of the nasal fossa, in front of the inferior turbinate, is often devoid of cilia. Immediately beneath this epithelium is a fibro-elastic tunica propria with a loose cellular subepithelial region. This contains numerous collections of lymphoid tissue which are particularly prominent in children. The tunica propria becomes more fibrous in its deeper layers becoming indistinguishable from the underlying periosteum and perichondrium.

The mucous membrane of the nose contains two types of secreting glands:

1. Goblet cells

These are unicellular glands contaning a nucleus and a clear, superficial area which represents the intracellular mucus which is discharged on to the surface of the epithelium.

2. Multicellular glands

The cells of these glands are like the goblet cells, but are collected together to line a pit or lumen. These compound glands may be simple tubes or more complicated branching structures.

The blood supply

The lateral wall is supplied by arteries derived from the spheno-palatine branch of the maxillary artery. This artery supplies a branch to the septum which also receives blood from its palatine branches as well as from the superior labial artery at the anterior end. Anterior and posterior ethmoidal branches of the ophthalmic artery also take part in the supply of the nose.

The blood from the nose drains with the sphenopalatine vein into the pterygoid plexus, into the facial vein in front and into the orbit with the ethmoidal veins. There are tiny veins passing through the cribriform plate which afford a connection with the veins on the orbital surface of the frontal lobe. This drainage is from a cavernous plexus deep to the mucous membrane and which is particularly well developed over the inferior and middle turbinates and the lower part of the septum.

There is now no doubt (Dawes and Prichard, 1953) that arterio-venous anastomoses exist in these areas. The arrangement is then as follows: the arterioles, which are initially arranged in parallel longitudinal rows, supply a capillary network just beneath the epithelium and around the mucus glands. This capillary system drains into the large venous sinuses which form the cavernous plexus. This plexus then drains into a deeper venuos plexus and from thence the blood is carried by venules into the main venous system. The remarkable features here are the arterior-venous anastomoses and the cavernous plexus. The walls of the latter are richly endowed with elastic fibres and circular and spiral plain muscle fibres. The muscle fibres are so thick in places that they can be shown to form sphincters in many places. This arrangement produces a very effective erectile system which can swell up, engorged with blood, and empty out again very rapidly.

Lymph drainage

The front part of the nasal cavity drains into the superficial lymph-channels which drain the skin of the face and end in the sub-mandibular lymph nodes. The rest of the cavity together with the sinuses drain into the deep cervical lymph nodes.

An important finding has been that the lymphatics of the nasal cavity can be injected from the subarachnoid space through channels which accompany the olfactory nerves. This makes it possible that there is a retrograde flow from the nose to the cerebrospinal fluid.

The nerve supply

The nerve supply of the nose is extremely complex and should be considered in detail because even the non-olfactory elements have a close relationship to the sense of smell.

The nose is supplied by nerve fibres from four sources:

1. Olfactory	—	Olfactory nerves. First cranial)
		Vomero-nasal nerve. (Accessory First)
2. Common sensation	—	Trigeminal. (Fifth cranial)
3. Sympathetic	—	Branches of the
4. Parasympathetic		sphenopalatine ganglion

Olfactory

The nervous connections of the olfactory organ will be described in the next chapter.

In many animals the vomero-nasal organ serves an olfactory function and it is connected by a special nerve to a small, accessory olfactory bulb. In the human, although a vomero-nasal pit is present in the 20-week fetus, it has usually disappeared in the adult. Occasionally a vestige of the organ is found, but no nervous connection or accessory bulb has been described.

Common sensation

Whenever the sense of smell is being investigated, it is particularly important to remember that the nasal mucosa can be stimulated also by touch, such as the impact of a stream of air; by cold such as that created by evaporating menthol; by pressure, heat and pain. These sensations are conveyed not by the olfactory system, which may be wholly deficient, but by the sensory fibres of the trigeminal, or fifth cranial nerve. They leave the nose in the anterior ethmoid branch of the nasociliary nerve and thus join the ophthalmic division of the trigeminal; but mainly they join the maxillary division through the anterior superior dental nerve, the long sphenopalatine nerve, the greater palatine nerve and nasal branches of the sphenopalatine ganglion.

These sensory fibres are the peripheral branches of the pseudo-unipolar cells of a semilunar ganglion which occupies a recess in the dura covering a shallow fossa on the apex of the petrous part of the temporal bone. The central fibres of these cells join the sensory root of the fifth nerve and enter the brain at the upper border of the ventral surface of the pons. The fibres end in the principal (superior) sensory nucleus of the nerve and in the nucleus of the spinal tract of the trigeminal nerve. The latter fibres are believed to convey mainly painful and thermal sensibility.

The activity of the vascular erectile tissue and of the secretory glands is dependent upon impulses conveyed by the sympathetic and parasympathetic outflows. It has been said that the 'natural' state depends on a 'balance' of these two sources of impulses, and that this 'balance' can be upset by interference from other exterior or interior impulses. This concept has led to the suggestion that various conditions are due to 'imbalance'. In the nose a condition involving excess mucous secretion and congestion of the erectile tissue often called 'vasomotor rhinitis', which is an adequate descriptive term, has been placed among these conditions and attempts have been made to control it by surgical interruption of the parasympathetic nerve fibres (Golding-Wood, 1961).

The sympathetic supply

These fibres originate as the medullated axons of connector-cells lying in the lateral horn of the grey matter in the first and second thoracic segments of the spinal cord. They leave the spinal cord in the anterior nerve roots of the corresponding spinal nerves and become separated from these nerves at the white rami communicantes to the corresponding ganglia on the sympathetic chain. The axons travel upwards to synapse in the superior cervical ganglion. The post ganglionic fibres form a plexus round the carotic artery and then rejoin inside the skull to form the deep petrosal nerve. This nerve combines with parasympathetic fibres to enter the pterygoid canal as the vidian nerve. When it emerges from the pterygoid canal into the pterygo-palatine fossa, the vidian nerve passes its sympathetic fibres to the sphenopalatine ganglion where they do not synapse, but continue in its nasal branches to the mucosa of the nasal cavity.

The parasympathetic supply

This originates in the superior secretory nucleus in the brain stem. The preganglionic fibres leave the brain as the *nervous intermedius* which forms part of the seventh cranial nerve, or facial nerve.

At a swollen bend on the facial nerve, known as the geniculate ganglion, the parasympathetic fibres leave the main nerve and form the greater superficial petrosal nerve. It is this nerve which joins the sympathetic fibres of the deep petrosal to form the vidian nerve. The parasympathetic fibres, however, synapse in the spheno-palatine ganglion before being distributed in the same manner to the nasal mucosa.

The paranasal sinuses

These are cavities around the nose, lined by a respiratory mucous membrane very similar to that of the nose itself. They are the frontal, ethmoidal and maxillary paired sinuses and the unpaired sphenoidal sinus. They communicate with the nose by small openings through which air is exchanged, and the mucus secreted into sinuses is wafted out by the lining cilia.

FUNCTION

The nose has two main functions:

1. It contains the organ of smell and its structure has therefore to be arranged in such a way as to permit odoriferous molecules to reach the receptor cells.

2. It provides an airway for respiration and this again implies a number of corollary functions. It is of considerable value to have the organ of smell imposed on to the respiratory airstream, as the movement of the air carries the odorous substances, and it can serve as part of the protective mechanism which the nose supplies to the lungs. The fact that the nose is also near the mouth

gives it an important role in distinguishing, appreciating and rejecting foods.

Most animals have a separate buccal and respiratory passage with a long epiglottis apposed to the palate. In man and the anthropoid apes this situation does not exist and buccal respiration is both possible and common. Although it is not entirely satisfactory, as it causes drying of the tissues in the pharynx, there is little evidence that it is damaging to the lungs themselves. Surgery, in the shape of tracheostomy and laryngectomy operations, has exposed the trachea itself directly to the airstream and although in the early stages humidification is essential if dangerous crusting is to be avoided, in time the tracheal mucosa appears to adapt itself to the new situation and very little added moisture is needed.

Although not entirely essential, the functions of the nose other than smell are important. They can be classified in the following manner:

1. Filtration.
2. Moistening and warming of air.
3. Speech.

Before examining these functions in more detail, it is necessary to understand the features of the airstream through the nose.

The nasal airstream

Paulsen made the observation in 1882 that during inspiration there formed a vertical jet of air in the anterior part of the nose. Since then there have been innumerable published accounts of the patterns formed by the air currents in the nose. Every conceivable pathway has in fact been described, but the experimental techniques used have so often been open to criticism that a lot of this work has been valueless. Virtually all recent thinking on these problems is based on the detailed and systematic descriptions given by Proetz (1953). He studied a large number of cadaver nasal cavities carefully built up to stimulate the unshrunken mucosa, and also perspex casts of the nose by passing smoke through them and observing the patterns formed.

He demonstrated that there were three invariables to the structure of the normal nose which were responsible for the basic respiratory pattern:

1. The overall shape of the chamber.
2. The relative positions of the entrance and exit.
3. Their relative sizes.

Proetz found that in the normal, variations of these three features were so minor that the patterns formed by the airstream were remarkably constant. Naturally the question of what is 'normal' arises, but it seems that only gross pathological deformities cause important changes. The common septal spurs and deviations, for instance, do not appear to interfere with the stream in any major way.

The direction of the anterior naris is complicated. In its lower third its axis lies in the sagittal plane and is oval in section. The direction then changes tending towards the septum at an angle of 20°. It then tends again towards the midline, thus hugging the septum.

Where the vestibule meets the nasal chamber there is a narrowing which has been called the *ostium internum*. This is the narrowest part of the nasal cavity and creates a most important situation. The air which is inspired is forced through this narrow ostium which then acts as a nozzle directing the jet almost vertically upwards towards the olfactory region. The inferior turbinate by swelling and decreasing in size may therefore act as a valve controlling the inlet.

The posterior choanae are larger than the anterior opening. Their height is between 24 and 33 mm and their width is 7 to 10 mm and 12 to 17 mm at the roof and floor respectively. This also has been expressed as an index, the choanal index:

$$\frac{\text{Transverse diameter} \times 100}{\text{Vertical diameter}}$$

and is said to be slightly larger in females.

The airstream, then, which in inspiration has been directed sharply upwards faces no impediment at the wide posterior choana and therefore spreads downwards in a gentle curve until it bends again towards the pharynx as the epiglottis causes this obstruction.

According to Proetz (1963) the expiratory currents show an unlimited range of variation. The laminar flow seen in inspiratory currents does not exist and eddies are present everywhere. Indeed the only constant and consistent pattern was a division of the stream by the posterior end of the middle turbinate into a septal and a meatal portion, as well as a large principal eddy (Figs. 1.4, 1.5, and 1.6).

The basic difference between the streams is the fact that the inspiratory one enters the nose through a narrow channel and is able to retain a smooth laminar flow because of diminished resistance at the exit. As the reverse occurs in expiration the stream tends to be broken up initially and any spur, minor deviation or swelling can only add to the disturbances in the currents. Proetz felt that so long as they were warm, moist and clean, the expiratory streams had little clinical significance but that it was notable that only expired air reaches the ostia of the sinuses.

There is now no doubt regarding the value of Proetz's work as this was the first comprehensive study of air currents in the nose. Many of his experiments were repeated by the author using a different technique. Water was allowed to flow through a trans-

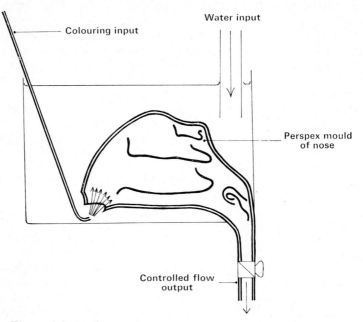

Figure 1.4   Technique for demonstrating the air currents in the nose using water.

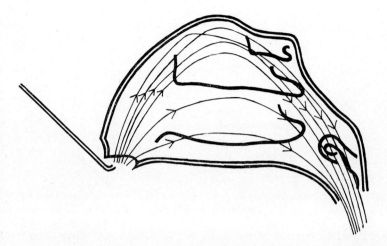

(a) Rapid flow (simulated sniff) showing laminar
flow mainly towards olfactory recess

(b) Slower flow (simulated quiet breathing)
showing flow mainly along floor of nose

Figure 1.5   The effect of flow rate on inspiration currents.

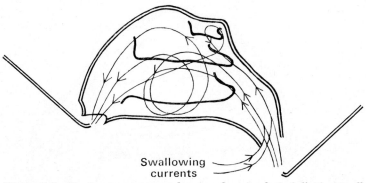

Swallowing
currents

Figure 1.6   Expiratory currents showing the mixed air full of virtually unrecordable eddies.

parent plastic replica of the walls of the nasal fossa, made from a plaster cast of the cadaver lumen.

The technique of using water instead of air is widely used in the engineering industry when the flow of gases has to be studied (Gerrard and McAreavey, 1957). It is particularly valuable when conditions are so complicated that calculations cannot be used to predict events. One of the considerable advantages is that, provided the right proportions of size and rates of flow are kept, the model can be scaled up or down. In these experiments, the size of the model was kept to the size of the nasal fossa in order to avoid adding to the calculations, but it is perfectly possible to upscale it in order to study more specifically certain parts or, using colorimetric methods, the *proportion* of gases reaching every area. This method of study may find some use not only in olfactory research but in the investigation of pathological conditions resulting from the inhalation of particles.

The rates of flow in water and air have to be proportional to the kinematic viscosities in order to get the same stream lines. The viscosity of water is 0·010 and that of air is 0·00018 poise. The ratio of the densities of water and air is 1 : 830 so that the ratio of the kinematic viscosities is 1 : 15. That means that the rate of flow of water has to be 15 times smaller than that of air in order to produce comparable currents.

As the volume of inhaled air in normal inspiration is about 250 ml for one nostril in one second, then the rate of flow of water should be about 16 ml per second.

During experiments a fine catheter delivering a dye, methylene blue, was placed at the opening of the model nostril and the resulting streams were observed in inspiration and in expiration and in simulated swallowing.

A number of findings should be recorded:

1. Inspiratory streams and laminar flow

Although the basic pattern described by Proetz held good it is incorrect to say that the flow was entirely laminar. Indeed in the region of the middle turbinate there was definite turbulence and numerous eddies.

It has been correctly stated that the best olfactory stimulus is obtained by placing the odorous substance at certain spots beneath the nostril and by sniffing. This has been used as argument against the presence of eddies in the inspiratory stream (Stuiver, 1958), the suggestion being that eddies would have brought odoriferous molecules up to the olfactory slit. As this is indeed the case, the argument is valueless. The reason why sniffing a suitably placed substance produces a better stimulus lies in the proportion of inspired air which reaches the olfactory area.

Another point which should be noted was that although most of the flow follows the curves demonstrated by Proetz, these flatten out considerably if the source of dye is placed in appropriate positions particularly if the rate of flow is slowed down. The result is a stream of air flowing along the floor of the nose which appears to end in the region of the eustachian opening .

During simulated sniffing a much larger quantity of the dye is sucked up towards the olfactory slit.

2. Expiratory streams and swallowing

As Proetz had demonstrated, expiratory streams broke into a constantly changing pattern of eddies. This turbulence allows the olfactory area to receive well-mixed portions of the expiratory stream which would contain some of the odoriferous molecules brought into the pharynx with the food.

During simulated swallowing small areas of turbulence can be seen bringing nasopharyngeal air forwards to mix with that in the nasal fossa.

The odour of food then is received from three sources:

1. The nostril, as food is brought towards the mouth.
2. During swallowing.
3. During expiration.

The rather diffuse manner in which the smell is perceived contributes to the nature of flavour and its distinctness from other smells.

Stuiver, using a stream of aluminium particles in a simple, flat-walled model nose, calculated that for normal inspiration between 5 and 10 per cent of the inspired air passes through the olfactory slit. At larger rates of flow this fraction becomes larger up to a maximum of 20 per cent. This larger rate of flow can be obtained by sniffing.

Pressures inside the nose

The movement of the thoracic cage sets up alternating positive and negative pressures in the nasal cavities. The resistance to be overcome in the cavity depends largely on its size, and the most important feature here is the relative sizes of the nostril and the posterior choana. The ostium internum, then, is not only responsible for the direction of the air currents but also for the intranasal pressures. The force and time needed for passing the 250 ml of air through a nostril depends on the cross-section of the nasal cavity at this narrow point. On average it is about 20 to 40 mm². In a broad nose with active alae nasi, movement of these structures alters the size of the cleft considerably, but in the narrow nose it is the swelling or shrinking of the conchae which controls the resistance of this area. For over a hundred years it has been known that the pressure inside the nose varies in inspiration and expiration by about $\pm$ 4 to 7 mm of water. Proetz has pointed out that this represents variations of only 1/2000 atmosphere.

Stoksted (1952, 1953) measured the degree of resistance in the nose and found it to be 18 mm of water, varying from 11·4 to 27. In disease, however, he found that it could vary from 2·3 mm to 1400 mm. Stoksted also noted that there was a rhythmic variation in the degree of resistance between one nasal cavity and the other. The total resistance, however, remained uniform. This bizarre rhythmical alternation in respiration from one nostril to the other has been called the nasal cycle and it has been demonstrated graphically countless times. Any person may note this in himself by careful observation and indeed a number present themselves to the doctor complaining of this symptom. Despite much speculation the cause of this phenomenon and the manner of its mediation remains unknown. Discussion of the nasal cycle usually leads to the observation that lying on one side causes obstruction of the undermost nostril and free breathing of the uppermost. This has been ascribed to a nasopulmonary reflex initiated by congestion and diminished aeration of the undermost lung. This theory is very attractive, particularly to those who like to reflect on the double nature of our paired organs, but not enough evidence has been produced to confirm it.

With these basic facts about the nose in mind, its functions can be considered.

1. Filtration

This means that the nose acts to some extent as a sieve trapping solid particles and preventing their entry into the larynx and the lower respiratory tract. The efficacy of this defence mechanism should not be overestimated as it is very rapidly overwhelmed in a dusty atmosphere. There is no doubt that workers in such conditions require protective masks.

There are four types of protection:

1. Hairs

A number of thick, coarse hairs also known as vibrissae grow in the nasal vestibule and arrest some of the coarser particles. The quanlity and length of these hairs varies considerably from person to person but are particularly prominent in men and in the elderly.

2. Cilia and mucus

The role of cilia and of mucus will be considered together as they are closely associated.

The whip-like movement of cilia has been observed so frequently in both the animal and vegetable kingdoms that their important place in the economy of living organisms has never been in doubt. In the nose and sinuses the cilia, carrying a 'blanket' of mucus have a cleansing function, wafting debris and bacteria backwards towards the nasopharynx. Cilia cover the lining of the nasal cavity except for the region just in front of the inferior turbinates and the small olfactory area. The sinuses are completely covered by cilia and they extend half-way down the nasopharynx where the epithelium becomes squamous in type. It is noteworthy that from that point the mucus can easily trickle into the pharynx where it is swallowed mixed with saliva.

These cilia are 5 to 10 μm long and 0·1 to 0·3 μm in diameter and appear to be protoplasmic rods with a denser core. Electron microscopy has allowed Bloom and Engström (1953) and Fawcett and Porter (1954) to describe their structure in considerable detail. The protoplasmic core is formed by nine pairs of filaments lying in a circle and a pair in the centre. This collection is surrounded by a membrane. The central pairs are arranged side by side and in fact are all in the same plane so that all cilia bend in the same direction which is at right angles to their axis. At the foot of each cilium is a basal corpuscle with rootlets penetrating the structure of the corresponding cell. The number of cilia carried by an average columnar cell of 5 μm in diameter can be, according to Fawcett, as many as 200.

The most remarkable feature of these structures is the integrated nature of the beats and the fact that there is a distribution of effort, some cilia resting while other move, so that a regular wave is produced. No form of nervous control has been shown to be responsible for this highly organized system of progression.

The cilia move at the rate of 1000 beats per minute and as they are covered by a layer of mucus, this substance is carried onwards in streams which follow fixed pathways. The mucus from the sinuses is directed towards the ostia and hence into the nose and all the nasal mucus is carried backwards into the nasopharynx.

Various factors affect the movement of the cilia. Moisture is essential, but water and even physiological saline (Negus, 1958) is incapable of maintaining the activity of excised ciliary mucosa. Gray (1928) considered magnesium, calcium sodium and potassium ions essential. In any case ciliary activity can be maintained almost indefinitely in Ringer or Tyrode solution. According to Krueger and Smith (1957, 1958 and 1959) positive ions caused a drop of 200 to 300 beats per minute while negative ions could restore the beat and even increase it by another 300 beats per minute.

Mucus obviously fulfils the conditions of moisture of a specific quality but its movement depends also on its viscosity and quantity, two factors which, as will be seen later, alter considerably in disease.

Only extremes in temperature offset the ciliary beat. Proetz found that greatest frequency occurred between 18°C and 33°C. Motion ceased below 7 to 12°C and above 43 to 44°C.

The optimum pH was 7·5 and the extremes were 6·4 to 8·5.

Drying of air passing through the nose remains one of the most important factors in practice, and Dalhamn (1956) showed that a relative humidity of 30 per cent stopped all action after 3 to 5 minutes.

The rate of flow of mucus in the nose has been studied by Ewert (1965) who directly observed the flow under magnification by a Zeiss diploscope. He found this on average to be 4·2 mm per minute at a relative humidity of 43·6 per cent. The most important finding is the dependence of the rate of flow on the wide variations in the relative humidity of the inspired air. Above 70 per cent no harmful drying effect is noticed. Between 40 per cent and 70 per cent the drying effect is partly compensated for by the secretions and evaporation from the mucosa, but below 40 per cent relative humidity the drying effect preponderates. Smokers have a lower flow rate mean, 3·6 mm per minute at 43·55 per cent relative humidity, and this difference becomes even greater at a lower relative humidity. An interesting fact which Ewert showed was the effect of shunting the airstream away from the nasal mucosa in patients subjected to tracheostomy. The mucus flow rate began to increase immediately, the rate of increase becoming greatest during the second and third postoperative weeks. After four or five weeks it had increased fivefold. This decreased again on decannulation, returning to normal within days, so that histological changes are unlikely to have taken place and favouring the assumption that evaporation, changes in viscosity and drying of the secretions play the greater part in this connection.

The very marked differences noted in smokers leaves one to speculate on the relative effects of the smoke exhaled acting directly on the nasal mucosa and those of nicotine acting on the autonomic system.

Nasal mucus consists of a glycoprotein, mucin, in a solution containing various other ions. According to Mélon (1967) the concentrations in normal mucus are the following:

Glycoprotein content is dependent linearly and inversely with the rate of secretion. It varies thus between 40 and 5 m.g^{-1} for rates of 1 to 3 mgcm^{-2} min^{-1}. Above that rate the mean is 6·5 mg.g^{-1} and this remains constant.

Inorganic components

Sodium	between 120·69—135·92	mean 127·70	μeq.g^{-1}
Chloride	128·93—148·99	138·82	
Potassium	12·27— 25·01	17·30	
Calcium	3·23— 6·77	4·88	

There are considerable alterations in these constituents in pathological conditions and, as will be discussed later, these changes may be associated with changes in the sense of smell.

Lucas and Douglas (1934) first noticed that mucus appeared to consist of two layers. The explanation for this may lie in some very interesting studies made on nasal mucus by Breuninger (1964). He was able to bring it into alternating states of gel and sol by shifting the pH towards the acid or the alkaline side respectively. The level at which the change took place was pH 7·5 to 7·6. Although it is not possible to measure the pH of the mucus as it is formed, it is assumed to be slightly alkaline. The high carbon dioxide content of the expired air is considered to change the pH towards the acid side. This brings the nascent mucus sol into a gel liberating as it does this a considerable amount of water which is used to humidify the air.

This sticky moving blanket of mucus traps particulate matter including bacteria and wafts them backwards to be swallowed and destroyed by gastric secretions. Negus felt that this could be but a minor role of mucus and that its main function was connected with fluid exchange.

3. Lysozyme

Nasal mucus shows mucolytic enzymatic activity. Lysozyme was described by Fleming, and it was believed to have an important bacteriostatic and to some extent bactericidal action. The importance of lysozymes in defence against bacteria remains unknown but there is no doubt that the more virulent pathogens remain quite insensitive to it.

4. Surface adsorption

Proetz (1953) took a great deal of trouble measuring the electrical surface charges in the nose by means of a galvanometer and came to the conclusion that particles adhered to the nasal surface in this way. Negus (19??) gave this theory considerable prominence in his book, but whether electrostatic or not in nature, some adsorption of particles certainly does take place.

2. Moistening and warming of the air

These two functions are so interdependent that they should be considered together.

Humidification of the air in the nose is necessary for a number of reasons. A moist medium is first of all necessary for olfaction. In order to carry out its respiratory function the alveolar epithelium has to be covered by a thin film of fluid; some of this may be produced by secretion and transudation in the tracheobronchial tree, as occurs in tracheostomized patients, but in normal persons the nose contributes considerably to the humidification of the air. Ciliary action cannot take place in a dry medium as has already been described and the defensive value of the mucus blanket disappears.

According to Proetz the nose moistens the inspired air to a relative humidity of 79 per cent, and with another 15 per cent added by the trachea, the relative humidity in the terminal bronchi is 94 per cent. Most other workers, however, state that at that level it is fully saturated at 100 per cent.

The quantity of water which can be carried by the atmosphere in the form of vapour is the absolute humidity. This quantity varies with the pressure so that it is greatest at sea-level, and also with the temperature. Warm air can carry a greater quantity of water without condensation than cold air, and cooling therefore results in droplet formation. The quantity of water which is actually present in the air given as a percentage of the maximum amount which can be carried at that particular temperature is the relative humidity.

The temperature in the nasopharynx is usually one or two degrees above 30°C when the room temperature is around 20°C. The inspired air becomes progressively warmer so that it is at body temperature in the trachea. This increasing warmth also means that more water vapour can be held by the air and condensation cannot occur. In expiration the opposite takes places so that the temperature at the nostril is again only a little over 30°C, and moisture then becomes available again. Negus attached great importance to this recovery and conservation of heat and moisture

B

by the nose during expiration and produced many arguments against the suggestion that the maxillo-turbinal system was designed for losing heat from the body.

3. Speech

The nose has two functions connected with speech. It is valuable both as a resonator and in articulation. Although there is virtually no evidence to support the belief that the paranasal sinuses have any function as resonators, blockage of nose by one or other cause is so common that the resulting 'nasal tone' is only too well known.

In the expression of words air has to be passed through the nose in order to articulate certain consonant sounds. When the nose is so obstructed that this is not possible the type of speech produced is referred to as rhinolalia clausa.

The opposite, rhinolalia aperta, occurs in cases of short palate or cleft palate. The nose cannot be closed off from the mouth and pharynx and correct production of vowels is not possible.

CHAPTER TWO

The Olfactory Organ and Its Connections

The anatomy of the olfactory organ is very similar in all animals from cyclostomes to mammals. This fact has led to the acceptance of the sense of smell as a very primitive system of nervous connections from which the rest of the brain has developed and to some extent at least this is likely.

The peripheral olfactory system consists of the olfactory organ—an area of specialized epithelium—together with the olfactory glands or Bowman's glands and the olfactory pigment. The olfactory bulb and its connections form the central part of the system.

THE PERIPHERAL OLFACTORY SYSTEM

1. The olfactory epithelium

In man the olfactory mucosa is limited to a region which includes the upper part of the superior turbinate, a corresponding area of the adjoining nasal septum and the roof in between. The margins of this organ are indistinct and irregular so that its exact area is difficult to estimate. There may also be a wide variation in size, as has been demonstrated in a number of animals. According to Brunn (1892) it is 10 cm² in man. This can be compared with 9·3 cm² in the rabbit (Allison and Warwick, 1949) and 169·46 cm² in the German shepherd (Lauruschkus, 1942). It is difficult to draw conclusions from these figures on the existence of a relationship between the area of the olfactory epithelium and the degree of development of the sense of smell as it is the number of receptors that really matters. In the rabbit Allison and Warwick estimated 100,000,000 receptors whereas in man the number has been estimated at ten times less.

The classification which we still use, recognizing that three types of cell form the epithelium, was first established in 1856 by Schultze. Electron microscopy, however, has added so many features to the known structure of the cells that the descriptions given here are based on recent micrographs.

The three types are: Receptor cells, supporting cells and basal cells. Deep to them, in the sub-epithelial tissues, are found specialized glands, Bowman's glands.

1. The receptor cell

This cell is a bipolar neuron, which has a cell body with two processes, a proximal and a distal one.

The cell body is round or oval, almost completely occupied by the nucleus. The nuclei of the receptor cells are all at a deeper level than those of the supporting cells which surround them.

The distal process is also known as the *rod*. It is 20 to 90 μm long in the rabbit and less than 1 μm wide (Le Gros Clark, 1956) and contains most of the cell cytoplasm. It is very rich in mitochrondria as well as microtubules which are mainly longitudinally orientated, and vesicles. There is a terminal swelling, or *knob*, at its peripheral end which is sometimes referred to as the 'olfactory vesicle' although such description would imply that it is hollow. This knob protrudes about 2 μm above the surface of the mucosa and bears 1 to 20 cilia which arise from basal bodies in the outer, more fibrillar cytoplasmic areas. There are also a number of centrioles which are however not paired with the basal bodies. The large quantity of mitochrondia in this region indicates a high level of metabolic activity which is not exceptional in sensory nerve endings.

THE CILIA

These fine structures project into the cavity which harbours the odoriferous molecules, so it has been generally assumed that the receptors are present on their surface membrane. There is, however, as yet no clear evidence of this. The length and motility of the cilia have been a matter of some controversy. Early reports suggested that there were short cilia which showed slow irregular movements and long cilia which did not move; it is more likely, however, that short cilia represent only broken immobile long ones and that their movement is brought about by the breakage.

The cilia consist of a shorter, thicker proximal segment, about 1 μm in length, which contains an array of nine pairs of subfibres (or ciliary tubules) plus two central subfibres. The distal segment is much narrower and tapering so that it contains only half that number of subfibres.

It has not been possible to obtain valid results for the length of the cilia although this is important, as it would indicate whether they remain enmeshed among the microvilli of the supporting cells, whether they bend along the air–liquid interface or whether they project into the layer of viscous secretion which is deep to the watery layer.

If it is accepted that the olfactory receptor sites are on the ciliary surface membrane then the most likely function of the cilia is to increase the total receptor-bearing area. The lack of motility makes it unlikely that they have a mechanical function in wafting odoriferous particles to and from the, as yet, undiscovered receptor sites.

THE PROXIMAL PROCESS

This is also long and thin, about 0·2 μm in diameter and is invested by the basal cells until it perforates the basement membrane where it immediately comes into association with Schwann cell cytoplasm.

These slender axons come together in the submucosa to form fasciculi which travel in a parallel to one another towards the openings in the cribriform plate of the ethmoid. Landau (1942) believed that these central processes formed a syncitial network in the deeper layers of the epithelium and therefore did not reach the olfactory bulb. The work which Le Gros Clark and Warwick (1946) carried out, and which demonstrated that retrograde degeneration of the olfactory bulb, make this view highly unlikely. It should be noted, however, in this context, that there is considerable close contact, not only between proximal processes but also between the cell bodies and distal processes of the receptor cells (Frisch, 1967) and that this contact may have some bearing on cell function.

2. The supporting cell

This cell is thicker than the receptor cells. It has a large oval nucleus which lies at a more superficial level than the nuclei of the receptors. Its distal process is very much thicker and ends in a surface of microvilli which branch out in all directions and fuse with other branches. This dense network extends beyond the receptor knobs and entraps the microvilli. Bloom and Engström (1953) have estimated that in man there are about 1000 microvilli for each cell.

The supporting cell has a much denser endoplasmic reticulum

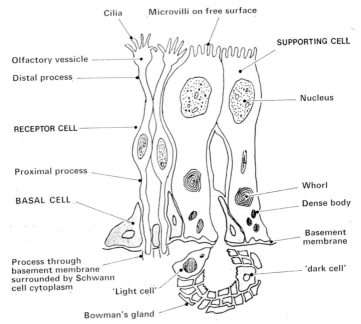

Figure 2.1   Ultrastructure of olfactory epithelium.

than the receptor cell and contains numerous granules and whorl-like systems which are usually associated with glycogen metabolism. Rees (1965) has described structures very similar to secretory droplets in the distal region of the cell.

The complexity of this cell makes it very unlikely that its sole function is to give mechanical support. The close manner in which its cytoplasm invests the receptor cell is analogous to the behaviour of the Schwann cells towards other neurons. Their contact is so intimate in places that the distance between the cells narrows to about 8 nm and the appearance suggests a fusion of the outer layers of the membranes (Rees, 1965). It is likely that at such places the barriers between the cellular constituents are much reduced so that some electrical coupling may take place. It has even been tentatively suggested (Tucker and Shibuya, 1965) that the electro-olfactogram is produced by the supporting cells when they are stimulated by the receptor cells.

Another aspect of these interesting cells is that in some species they have been seen to assume phagocytic properties when the adjacent receptors have been injured (Le Gros Clark, 1956). Finally, the secretory-like granules previously mentioned have led to the suggestion that the supporting cells are responsible for some of the secretions bathing the olfactory area; while others have pointed out that villous surfaces are often associated with absorptions (Fig. 2.1).

3. The basal cells

These are small, polygonal cells which lie on the basement membrane. They surround the axons of the receptor cells as soon as these lose their supporting cell envelope. The basal cell membranes are in close contact with the membrane Schwann cells which protect the axons when these cross the basement membrane into the subepithelial tissues (Fig. 2.2).

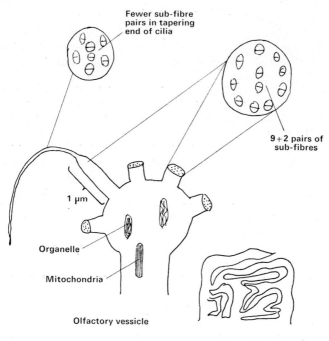

Fewer sub-fibre pairs in tapering end of cilia

9 + 2 pairs of sub-fibres

1 µm

Organelle

Mitochondria

Olfactory vessicle

Dense net-work of microvilli on surface of supporting cell

Figure 2.2   Surface of receptor and supporting cells.

The basal cell nucleus is central, ovoid and dark, containing large clumps of chromatin with other dark particles in the inter-chromatin areas. The cytoplasm which contains mitochrondria, Golgi complex and granular endoplasmic reticulum, is also darkened by the presence of many free ribosomes which tend to form rosettes.

2. Bowman's glands

These glands have a considerable species variation. In man they have been described as tubulo-alveolar, while in a recent study of the ultrastructure of the mouse olfactory mucosa Frisch (1967) could find no regular alveolar structure. He could distinguish two types of cell, dark and light, according to their electron-density; both, however, may share the lumen of the same intercellular or acinar secretory canaliculus. The luminal surfaces are not smooth, but covered with microvilli and the cells contain numerous vesicles and secretory granules. This confirms their secretory function although microvilli are often associated with absorption.

It has not been easy to establish morphologically or histochemic-ally whether these are serous or mucous glands. Indeed, they show some features of both.

According to Larsell they are innervated by the nervus termin-alis but there has been no confirmatory work on this subject. The density of innervation, however, varies with the species and is richest in the dog.

The function of these glands, let alone their product, is not known. That they are connected with smell in some way has been deduced from the fact that they are found only beneath the olfac-tory mucosa. According to Negus (1958) glands of this type were present in all the animals which he had studied except fish, which only had the goblet glands of the respiratory musoca. This would lead to speculation as to whether the secretions of Bowman's glands are replaced by the constant washing of the olfactory area by water. It is unlikely that the fluid covering of this region is solely derived from Bowman's glands, as the mucus sheet originating in the respiratory mucosa of the nose and sinuses is also involved and possibly some contribution is made by the supporting cells.

It may be that these secretions are responsible for removing the odorous molecules which have stimulated the receptor sites. The author and M. Poynder (1968) have shown the influence of nasal mucus on the perception of smell; it is possible that the secretions of Bowman's glands have a similar adsorptive effect. It is worth noting that of all the animals studied the one with the most developed glands is the Egyptian vulture *Neophron perinopterus* (Bang, 1964). This bird feeds chiefly on excrement.

3. The olfactory pigment

It had long been noted that in many vertebrates the olfactory mucosa has a yellowish or brownish colour. This had led to the assumption that the pigment involved had something to do with smell. This belief was further reinforced by the importance of pigment in visual perception and a comparison was often made with the yellow pigment which is present in the deeper layers of the retina, at the fovea centralis.

Although the similarity with cutaneous pigmentation is also frequently quoted, and there is some evidence (Kingsley, 1919) that in fish and amphibia cutaneous melanophores become included in the ectoderm from which the olfactory mucosa originates, such cells have never been seen in the vertebrate epithelium or submucosa.

Moncrief believes that this pigment is important in olfaction. He refers to the black pigs which are bred by Virginia farmers in preference to white pigs. The latter die from eating poisonous plants, *Lachantes tinctoria*, because, it is thought, they have a poor sense of smell and are unable to recognize them. He gives as examples also the white sheep of the Trentino in Italy which eat and are poisoned by St John's Wort, and the Himalayan rabbits that are born white and acquire pigment only when weaned.

Albinos, in the same manner, have been said to be congenitally anosmatic (Mackenzie, 1923).

Most of these examples of albino animals being poisoned by plants were given by Darwin in 1868, but he himself never assumed that this was due to reduced olfactory powers. Darwin concluded that all the animals eat the plants but only the pigmented ones survive. Moulton and Biedler (1967) suggested that these examples are instances of photodynamic sensitization. Moulton (1960) found that of rats trained to recognize three aliphatic alcohols, albinos had lower thresholds than their pigmented littermates. A link between albinism and hyposmia implies that the olfactory area of albinos is deficient in pigment. No evidence of this has been produced and Moulton (1962) showed that the mucosa of albino and pigmented rats was indistinguishable either in appearance or in chromatography of the extracts. This has been confirmed in other species (Jackson, 1960).

The pigment itself is present in the form of granules in Bowman's glands as well as the supporting cells. Most workers have been unable to show the presence of these granules in the receptor cells themselves.

There are two types of yellow pigment in olfactory mucosa, carotenoids and non-carotenoids.

1. Carotenoids

Many workers have shown the presence of both free and protein-bound carotenoids as well as free vitamin A in cow mucosa. Negative results however have been obtained in pig, salmon and rabbit mucosa. This does not mean that pigment in very low concentrations would not have escaped detection, but it does prevent a firm acceptance of the role of carotenoids or vitamin A in olfaction.

2. Non-carotenoids

These are associated with lipids and are of an extremely complex nature. The lipids are normally phospholipids, mainly lecithius, and Moulton (1962) was able to separate 4 fractions from the saponifiable fraction of an extract of cow mucosa using paper chromatography. The associated pigment was found to give negative results when tested for the common body pigments such as melanin. Jackson (1960), however, came to the conclusion that the pigment was partly the phospholipid but that most of the colour was due to the products of auto-oxidation of these lipids.

4. The olfactory nerves

These are formed from the proximal processes of the receptor cells and are non-medullated, being surrounded by a nucleated sheath. As the fibres cross each other in the submucosa to form bundles they give an appearance of a plexiform network. These bundles group themselves into two, one group of between six and 20 entering the skull by the foramina in the lateral part of the cribriform plate and the second group of somewhat fewer bundles crossing the medial parts.

They immediately enter the olfactory bulb and end in its glomeruli.

Each of these bundles is ensheathed by a projection of pia-arachnoid, while the dura becomes continuous with the periosteum of the nasal fossa; in this way the subarachnoid space communicates with the perineural spaces of these nerves.

THE CENTRAL OLFACTORY SYSTEM

1. The olfactory bulb

In a number of animals the olfactory bulb is a hollow structure which communicates directly with the lateral ventricle through a hollow stalk—the olfactory tract. In man the walls of the bulb cavity, especially the ventral wall, have become so thick that the cavity is either obliterated or filled with neuroglia.

In appearance it is oval, similar to brain tissue. It lies on the orbital plate of the frontal bone immediately adjacent to the cribriform plate of the ethmoid. Through this plate the olfactory nerves pass and enter the vertical surface of the bulb. Immediately above it is the frontal lobe and it lies beneath the anterior part of its olfactory sulcus (Fig. 2.3).

Its posterior surface is continuous with a thick band of fibres—the olfactory tract which lies along the olfactory sulcus. Where this tract enters the brain it divides into two roots, lateral and medial, so that the triangular area which these enclose is called the *olfactory pyramid*. This area is continuous posteriorly with a region referred to as the *anterior perforated substance*, because it is pierced by

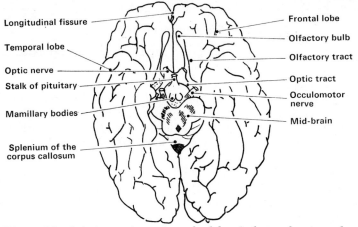

Figure 2.3   Inferior surface of cerebral hemispheres showing relations of the olfactory bulbs and tracts.

many blood vessels. It is the part of the basal surface of the brain which lies between the optic tract and the uncus. Sometimes it is possible to see a small elevation just behind and lateral to the pyramid; this is called the *olfactory tubercule* (Fig. 2.4).

When the fasciculae which are formed from the olfactory nerves enter the anterior cranial fossa they become interwoven in an extremely complex fashion, and it is this thick mesh which forms the outermost layer of the olfactory blub. It has been impossible to demonstrate any topographical localization of these fibres using

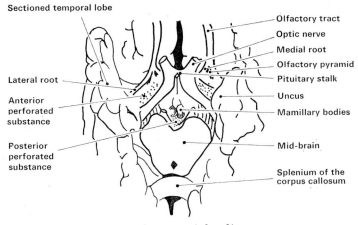

Figure 2.4   The roots of the olfactory tract.

histological techniques, but some results were obtained using experimental methods. Adrian (1950) stimulated small areas of the olfactory epithelium of the rabbit and recorded the response in different parts of the bulb. He found that the oral extremity of the epithelium was related to the anterior part of the blub and the aboral extremity to the posterior part. Le Gros Clark had already shown in some important work carried out with Warwick (1946) that extirpation of the bulb caused retrograde degeneration of receptor cells. He later (1951) inflicted partial lesions of the bulb, and found that its lower part was related to the upper areas of the epithelium. This was less definite, however, in the antero-posterior plane.

In other words there is clearly some partial localization though not as clear cut as between the retina and the geniculate nucleus.

MICROSCOPIC STRUCTURE

The dense, white interlacing fibres which form the surface of the bulb end in little spherical masses of dense nervous tissue. These are called olfactory glomeruli and consist of the synapses between the primary olfactory neurones and the dendrites of certain mitral and tufted cells. The axons of these two types of cells form projection pathways to the secondary centres of the fore-brain.

Numerous short axons are also present. They are collaterals of the mitral-cell and tufted-cell axons and they probably serve to relate several glomeruli. The neuronal chains which they provide are probably responsible for the spontaneous activity observed in electrical recordings from the olfactory bulbs.

In mammals these features result in a distinct lamination where six layers can be recognized (Fig. 2.5).

1. Olfactory nerve fibres.
2. Glomeruli.
3. Outer plexiform layer. This consists of:

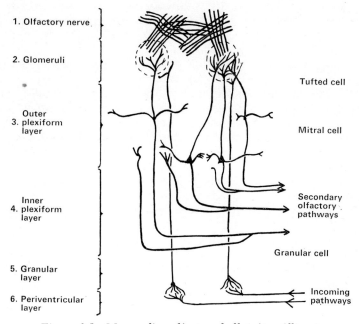

Figure 2.5   Mammalian olfactory bulb (after Allison).

(i) Mitral cells. These are large and have only one main apical dendrite which enters a single glomerulus, and a number of accessory dendrites which end in this layer.

(ii) Tufted cells. These are smaller than the mitral cells but have a similar arrangement of dendrites.

4. Inner plexiform layer. This layer consists of the axons of the two cells in the previous layer, together with a large number of recurrent dendrites.

5. Granular layer. This contains a third or granular type of cell. These cells are surrounded by a plexus of fibres of mainly extrabulbar origin, while their axons end in the outer plexiform layer around the dendrites of the mitral and tufted cells. This apparently provides a pathway for central influence on to the olfactory bulb.

6. Periventricular layer. These white fibres are mainly extrabulbar in origin.

THE CONNECTIONS OF THE OLFACTORY BULB

According to Allison (1953) there are two relatively independent pathways which relay olfactory impulses from the glomeruli.

1. Originating in the mitral cells

These fibres form the *lateral olfactory tract system* and end in superficial secondary centres, namely:

(i) The anterior olfactory nucleus—mainly its pars lateralis. This nucleus is present in all vertebrates and is immediately behind the bulb. In mammals four parts have been recognized:

A dorsal part continuous with the neopallial cortex,

A medial part related to the anterior continuation of the hippocampus,

A ventral part which usually joins the deep layers of the olfactory tubercle,

A lateral part which continues postero-laterally into the prepyriform cortex;

(ii) The prepyriform cortex.

(iii) The periamygdaloid cortex.

(iv) The olfactory tubercle which is part of the anterior perforated substance.

(v) The nucleus of the lateral olfactory tract.

(vi) The cortico-medial group of nuclei which form the amygdaloid complex.

2. *Originating in the tufted cells*

These form the medial fibres continuing into the anterior commissure, and end in more medial centres:

(i) The central amygdaloid nucleus.

(ii) The nucleus of the stria terminalis.

(iii) The opposite bulb via the anterior commissure.

Lohman and Lammers (1963) disagree with this view and say that the lateral olfactory tract is the only projection bundle.

Brodal (1963) says that the vast majority of fibres which leave the bulb as the discrete lateral and more diffuse medial and intermediate olfactory striae end in the anterior olfactory nucleus. Very few fibres pass it by. A third tier of fibres then travel centrally to the:

(i) Pyriform lobe cortex (the gyrus hippocampi in man);

(ii) The nucleus of the lateral olfactory tract, which is, however, less well developed in man;

(iii) The amygdaloid complex:

(iv) The olfactory tubercule and cia the tuberculo-septal tract to

(v) The septal nuclei;

(vi) The diagonal band of Broca.

Brodal suggests that there may be no direct commissural fibres between the two bulbs, but that communication is by fibres originating and ending in the anterior olfactory nuclei.

2. The 'olfactory brain'

Although at the present time there is hardly a part of the brain more closely studied than the *rhinencephalon* or 'olfactory brain'; it is not so long ago that it was the most neglected area. This general lack of interest was probably mainly due to the fact that all the structures which had been pushed on to the medial and undersurface of the cerebrum by the great development of the neopallium in man, had been assumed to be connected 'with smell'. This area is furthermore not easily approached by either experimentalist or surgeon and the electrical discharges it generates are far from the commonly used electrodes.

It was the anatomist Papez who more than 20 years ago first suggested that this so-called 'olfactory brain' was more concerned with affective behaviour than with smell. The term rhinencephalon would then hardly apply to most of these structures, and MacLean (1952) suggested the name *limbic system* derived from Broca's *great limbic lobe*. To these paleocortical structures in the medial

and basal parts of the cerebrum should be added the closely related subcortical amygdaloid complex.

The final connections and pathways of the olfactory system are
still largely hypothetical. Earlier beliefs that the 'olfactory area'
of the cerebral cortex lay in the hippocampus have little evidence
to support them. From a morphological point of view the hippocampus is an effector area and not necessarily connected with reception or association. Study of its afferents implies that the
majority of these fibres do not reach the hippocampus directly, but
are relayed from the prepyriform and periamygdaloid areas (in
man these are represented by the gyrus ambiens and semilunaris
respectively). The fornix is the main, and probably the only,
afferent bundle of fibres from the hippocampus. Most of the fibres
lead to the mammillary body, mainly on the homolateral side but
others lead to the hypothalamus, as well as partly crossed fibres
to the septal nuclei, the pre-optic area and the habenula.

Fibres from the mammillary body are relayed to the anterior
thalamic nucleus and from thence to the cingular gyrus. Those from
the hypothalamus are relayed to the frontal lobe or join septal and
pre-optic fibres in the stria medullaris to form a descending pathway to the brain stem.

Fibres also reach the hypothalamus, septal and pre-optic nuclei
from the amygdaloid complex via the stria terminalis.

The entorrhinal area which is equivalent to the gyrus ambiens
and the anterior part of the gyrus hippocampi can be classed with
the secondary optic, acoustic and somaesthetic cortices. In other
words it is mainly concerned with association and integration of
olfactory impulses.

The amygdaloid complex contains two separate cell masses
which form the olfactory part, a central nucleus which has olfactory projections by way of the commissural projection system and
the cortico-medial nucleus by way of the olfactory tract.

Allen (1941) found that bilateral ablation of the pyriform and
amygdaloid areas in dogs abolishes complex olfactory conditioned
reflexes. The ability to carry out simple olfactory discrimination,
however, is unimpaired unless the subcortical olfactory areas are
also damaged.

FINAL CONNECTIONS

In fishes, amphibians and reptiles there are two main projection
pathways, the olfacto-hypothalamic tracts which relate the basal

olfactory areas to the visceral hypothalamic areas, and the olfacto-habenular tracts, through which tertiary fibres pass to the habe-nular nuclei and thence through the habenulo-peduncular tract to the motor region of the tegmentum.

In mammals the further connections of the olfactory system are also difficult to investigate. Secondary centres such as the amygdala have even more important non-olfactory connections and the only area which appears to be specifically olfactory, the prepyriform cortex, is not easily accessible for experiment.

Allen (1943b) stimulated the prepyriform area in dogs and re-corded spikes about 3 milliseconds later from the centro-lateral portion of the *prefrontal* area only. Undercutting this area abolished the responses.

Allison (1950) destroyed the prepyriform cortex in the rabbit and traced the degenerating fibres into a part of the granular insu-lar cortex which was equivalent to the olfactory area described by Allen, on both sides of the brain. The uncrossed fibres pass through the uncinate fasciculus and external capsule; the crossed fibres through the anterior commissure.

In other words the olfactory sense, like the other senses, is finally represented in the isocortex. It should be noted that taste also has an isocortical localization in the parinsular region near the area for somatic sensation from the tongue. Thus smell and taste areas lie very near to each other and near the region from which mastication and salivation have been elicited by electrical stimulation.

Other fibres from the prepyriform cortex and the amygdala travel through the stria terminalis to the medial pre-optic area. Therefore, there are two main fibre projections:

1. *A cortical termination* mainly in the *prepyriform area*, but also in the dorsal and external parts of the *anterior olfactory nucleus*, in part of the *olfactory tubercle* and the superficial nuclei of the cortico–medial area of the *amygdala*.

2. *A sub-cortical termination* in the bed of the *stria terminalis* and in the *central amygdaloid group*.

This may mean that there are two distinct levels of olfactory activity:

1. A more refined discrimination associated with intelligent re-actions;

2. A crude discrimination associated with more automatic reactions.

A final comment should be made on the 'rhinencephalon'. Its constituent parts are usually considered to be the following Gray's Anatomy, 35th edition):

1. Olfactory bulb.
2. Olfactory tract.
3. Olfactory pyramid.
4. Olfactory tubercle, prepyriform.
5. Olfactory anterior perforated substance prepyriform.
6. Olfactory pyriform area.
7. Olfactory hippocampal formation.
8. Olfactory paraterminal gyrus.
9. Olfactory fornix.
10. Olfactory nucleus habenulae.

In view of what has been demonstrated regarding the function of the hippocampus, its connections, and the role of the amygdala, this list does not appear very useful. It could be said that one smells with the whole brain, or that only the direct sequence of neurons starting at the olfactory bulb should be included, leaving out all the associated areas. Because of the difficulties in deciding what to retain and what to leave out and because of the largely hypothetical nature of the further connections of the olfactory system, it may be as well to retain the term rhinencephalon in its traditional sense.

CHAPTER THREE

The Electrophysiology of the Olfactory System

The olfactory receptor cell is able to detect minute quantities of certain chemicals which are recognized as odorous substances. The stimulus which these impart is transduced into a coded message that provides the brain with information about the quality and quantity of the stimulating agent.

The receptor cell is a bipolar neuron and experimental evidence suggests that the primary reaction with the stimulating agent takes place in the membrane of the hairs. We still do not know the essential stimulus and research has not advanced us beyond the realm of hypothesis. The various theories relating to this subject will be discussed in Chapter 4, but it is preferable first to examine the electrophysiological evidence available associated with each element of the olfactory system.

THE OLFACTORY MUCOSA

The electro-olfactogram

The greater part of the work on the electrical responses of the olfactory mucosa has been carried out by Ottoson from 1956 onwards. He placed a ringer-agar electrode on the surface of the olfactory mucosa of a frog and a puff of odorized air was directed towards that region. A slow, negative, purely monophasic potential was developed (Fig. 3.1). By analogy with the electro-retinogram Ottoson has called this the electro-olfactogram, or e.o.g.

Most of this work has been carried out in the frog because of the simplicity of its olfactory organ and the relative ease with which an electrode can explore its surface, but studies in the rabbit and guinea-pig have shown that there is no essential difference in the response in those species. Indeed even insect olfactory receptors have shown similar potential changes. This response is not obtained in parts of the nasal mucosa which are not anatomically

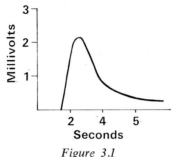

Figure 3.1
An electro-olfactogram.

part of the olfactory region and are not produced by the mechanical stimulation of non-odorized air.

In the frog the response varies from one region to another, the greatest potential being obtained from the areas where the receptors are most numerous anatomically.

Cocaine studies were used to prove that the potential comes from the receptor cells and not the nerve fibres as cocaine blocks these fibres but does not interefer with the e.o.g.

The relationship between the e.o.g. and a possible 'generator potential' for the transmitted nerve impulse is not clear. The e.o.g. may be a composite potential of which the generator potential may be a part together with others, possibly even of non-receptor origin. No e.o.g. has been convincingly demonstrated in man.

Although the e.o.g. is a mass response and gives little information about the function of the individual receptors which are very difficult to isolate, it can be used as an index of activity of the olfactory membrane. The effects of different substances can then be measured quantitatively.

The temporal course of the discharge

It is a well known fact that the olfactory sensations produced by certain substances are more persistent than those of others. The e.o.g. responses to various substances can be compared provided that the stimuli are of equal intensity and this shows a considerable variation in duration (Fig. 3.2). Thus amylacetate gives a fast-rising phase and a comparatively short-lasting falling phase while oil of cloves has a more prolonged fall.

It should be noted that compared with corresponding potentials in other sense organs the response of the olfactory membrane is extremely slow. In the frog a puff of air containing a low concentration of odorant and lasting about one second produces a response lasting 4 to 6 seconds. This would not be surprising if it is assumed that the odorous particles have to pass through a layer of mucus before reaching the membrane. Excitation would then take place gradually, its time course being dependent on the number of particles which reach the receptor per time unit. The time course of this excitatory process must also depend on the time it takes for the particles to become inactivated, or to cease to stimulate.

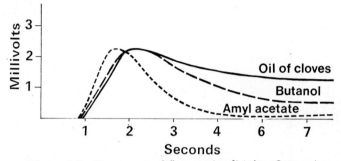

Figure 3.2   Response to different stimuli (after Ottoson).

The difference in time course of the response to different substances which Ottoson has noted raises extremely important questions. The full relationship between this and the physical and chemical properties of the odorant have yet to be worked out, but it is possible that the solubility properties of the stimulating substance and the vapour pressure are of decisive importance. The nature of the mucus and its composition must also be of some importance and the flow rate of the air current may have a role.

In this context it is interesting to recall that Adrian (1951) showed that the same difference existed in the time course of the electrical discharges in the secondary pathways of the bulb. As these differences are carried to the olfactory cortex it has been suggested that discrimination between the different substances may depend, therefore, on the temporal course of the initial discharge.

The amplitude of the discharge
Ottoson measured the increase in amplitude of the response with

increasing stimulus intensity, so that he could plot the relationship between the two (Fig. 3.3). This response appears to increase more or less logarithmically with increasing odour intensity. This finding has been rendered very interesting in view of the psycho-physical work of Engen ((1961, 1962). His studies were concerned with methods of scaling the intensity of odours and he demonstrated that the logarithmic values of subjective ratios increase linearly with the logarithm of the concentrations of the odorants.

Adaptation
Although it has been believed for a long time that olfactory

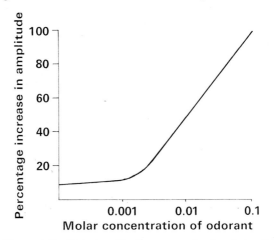

Figure 3.3   Relationship between stimulus intensity and amplitude of electro-olfactogram (after Ottoson).

receptors adapt rapidly and indeed this view still seems to be wide-spread, all the electrophysiological evidence is against it.

If a continuous stream of air containing an odorant is passed over the olfactory mucosa of an Ottoson frog preparation a sustained potential is produced. The level of this plateau is maintained with only very little decline, losing less than 40 per cent of the height of the initial peak. This indicates, of course, that the olfactory end-organ is among the slowly adapting sensors. Furthermore, if the olfactory mucosa is repeatedly stimulated the response declines only very slowly, retaining its amplitude for a considerable time.

In view of the manner in which the sensation of a smell rapidly weakens and becomes imperceptible the phenomenon of adaptation remains difficult to explain. Adrian, who showed that the response of repeated stimulation as measured in the bulb also retained its strength, suggested that the incoming signals might be suppressed by the intrinsic activity of the bulb, but there is a possibility that the efferent fibres of the olfactory system may be connected with this.

It is perhaps worth repeating that these mucosal responses represent the summated activity of all excited receptors and not the electrical events in a single receptor.

Single-unit activity

The difficulties involved in recording from a single receptor unit were not surmounted until Gesteland (1956) was able to study the response of single olfactory receptors by recording from the axonal extension of these cells in the frog. Although in earlier reports it was suggested that these units could be classified into a few groups, it is now believed that no two receptors respond in the same manner.

Using a saline electrode applied gently to the surface of the mucus layer and a metal electrode into the mucosa, Gesteland and his colleagues recorded some very interesting results.

The metal electrode was pushed through the olfactory mucosa, past the basement membrane until they could record the potentials in the receptor-cell axons. These took the form of triphasic, diphasic or monophasic spikes of some 200 to 300 μV peak-to-peak. Having established the basic response, they then stimulated the olfactory mucosa with various different odorous chemicals and the main findings they recorded were as follows:

(1) Some units are *excited* by some odours and *inhibited* by others.
(2) If two odours excite one unit equally, they need not have the same action on another unit. The same applies for inhibition.
(3) Stopping an odorous stimulus may itself have an effect on the discharge. An odour may produce a delayed response which may be either excitatory, or inhibitory or simply produce a more rhythmic firing pattern.
(4) Whenever they have taken two similar odours, such as menthol and methone, or nitrobenzene and benzonitrile and

have found units which respond similarly to both, a short search discovers a unit that discriminates between the two.

These findings, although not entirely conclusive, may be of considerable importance in understanding olfactory discrimination.

THE OLFACTORY NERVE

Because of their inaccessibility and the short length of the fibres it is very difficult to investigate the electrical activity of mammalian olfactory nerves. Ever since Garten recorded the action potential of the pike olfactory nerve in 1900, most of our information has come from amphibia and fish. Garten's preparation was itself used by Gasser (1956) to show that the action potential of the olfactory nerve consists of one single wave with a conduction velocity of 0·2 millisecond. This is slow and conforms with the finding that the axons in the nerve have an average diameter of 0·2 μm.

Ottoson (1959) showed that the action potential in frog has exactly the same simple configuration and that in this case the conduction velocity is 0·14 millisecond. This similarity is not surprising as all these fibres have approximately the same diameter.

The olfactory nerves are among the slowest conducting afferent systems in the body, but it should be remembered that the distance they travel is extremely short, so that in the frog they reach the bulb in 50 milliseconds. At any rate these findings suggest that time is of relatively little importance in the transmission of olfactory information to the brain.

In 1955 Beidler and Tucker developed a technique for measuring the olfactory nerve response to odour stimulation. They dissected a small bundle of nerve fibres which must have contained a minimum of 2000 fibres and pinched it off where it enters the bulb raising it on to platinum electrodes under mineral oil.

These same elctrodes can be used either to stimulate the nerve electrically or to record the neural discharges. By electrically and therefore antidromically stimulating this twig the arrival of the impulses can be recorded at the mucosa and the area they serve mapped out. It is found that twigs of 10 to 40 μm in diameter project to roughly oval areas of mucosa which are less than 1 mm². When the area is known so precisely, it can be stimulated more directly with odorous substances.

As can be expected, the record of the neural discharges which result from odorous stimulation is a barrage of overlapping impulses. This has been given the title of *asynchronous activity*, when all it means is that the record represents scrambled responses. However, using special electronic filtering equipment, it is possible to 'integrate' or 'summate' these records. The result is a running average of response in which the phasic components can be demonstrated (Fig. 3.4).

Mozell (1944) obtained recordings from two widely separated branches of the olfactory nerve of the frog. He did this by carefully dissecting out the most medial and the most lateral branches he could distinguish. A series of odours were blown onto the mucosa and recordings taken from both branches from which the maximum response obtained for each odour was noted. The ratio

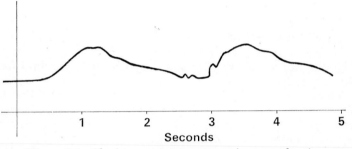

Seconds

*Figure 3.4   Phasic components average from overlapping
impulses.*

of the maximum response of the lateral nerve to the maximum response for the medial nerve was then worked out for each substance and this produced the following results:

1. Each substance elicited a different ratio.

2. The time lapse between the two nerve responses differed.

This suggests that the input to the central nervous system from the olfactory peripheral system differs in temporal and spatial discharge pattern for different chemicals.

One fact regarding these olfactory nerves may be very important. If the axons are so tightly packed within a Schwann cell tube, there is little more than a single cell membrane between them. This would be only 100 to 150 Å distance making separate firing of individual axons unlikely. Interaction between different and neighbouring receptor units is then quite possible.

THE OLFACTORY BULB

Intrinsic activity

Gerrard and Young (1937) first described the persistent activity which took place in the bulb in the absence of manifest stimulation.

In recordings from the surface the spontaneous activity appears as irregular waves which are accompanied by a discharge of impulses in the deeper layers.

This persists after section of the olfactory nerves or complete destruction of the olfactory mucosa. Section of its connection to the brain may even enhance its activity.

Induced activity

When the mucosa is stimulated with odorized air, three main types of response occur:

1. A slow sustained potential;
2. Oscillatory waves;
3. Impulse activity in the form of spikes.

The first two are observable from the surface and the third from the deeper layers.

There is some experimental evidence which indicates that the slow potential originates in the dentritic network of the glomeruli and that it is of the same nature as the sustained potentials recorded from the sensory cortical areas.

The second type of response, the oscillatory waves, which in the frog have a frequency of 8 to 12 per second are superimposed on to the slow potential. The frequency of these bursts of oscillatory activity differs markedly between species, being 40 to 85 per second in the rabbit and 20 to 40 in dog and monkey.

Some concentrated odours, like cigarette smoke, may produce a train of such waves whose amplitude rises to 3 MV or higher and which may last several seconds. Antidromic stimulation of the secondary neurons blocks these induced waves indicating their connection with these secondary pathways.

By simultaneously recording the e.o.g. in the frog's mucosa and the discharges in the bulb Ottoson (1959) was able to study the relationship between the events in these two areas (Fig. 3.5). The changes in stimulus strength which produce e.o.g.s of different amplitude, show identical changes in the amplitude and time course of the slow response of the bulb.

As the bulb-discharge can be regarded as a generator potential

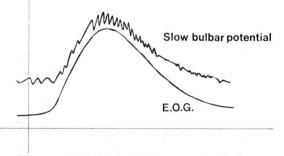

*Figure 3.5   Relationship between slow bulbar
potential and e.o.g. responses to butanol (after
Ottoson).*

linking the incoming signals with the olfactory cortex it is of particular interest that the events in the sense organ are faithfully reproduced at the first synapse.

The sequence of events may be as follows:

When the olfactory particles make contact with the receptors a change takes place in the receptor membrane creating a receptor potential. This potential sets off an afferent discharge along the olfactory nerve and this induces the slow synaptic potential in the glomeruli of the bulb. This closely follows the pattern of the receptor potential and induces a discharge in the neurons of secondary order.

As the stimulus intensity is increased, more and more bulbar neurons are thrown into synchronous activity, which is reflected as regular waves superimposed upon the slow potential change.

If an electrode is implanted into the core of the bulb of a rabbit at a depth of 1·5 mm a series of spike potentials are obtained at each inspiration of room air. These can be made more intense by adding an odorous substance. Adrian (1956) has shown that these spikes are very likely to come from the mitral cells or their axons. He pointed out that as in the rabbit there are only about 60,000 mitral cells to deal with the signals from 50 million receptors, then each spike picked up from a mitral cell axon represents the activity of 800 receptors. When one considers that the mitral cells are also not only connected with one another but with intra-bulbar feed-back loops and with efferents from higher centres in the brain, the spike becomes a most complex product.

Electrical stimulation of the receptors

An electric shock of 0·2 millisecond duration to the olfactory mucosa of the frog produces in the bulb a purely monophasic negative potential lasting 150 milliseconds. This response is built up from two components with distinctly different properties:

First component—which lacks a refractory period and summates to a sustained potential under repetitive stimulation. This component is probably synaptic in origin and corresponds to the slow bulbar potential.

Second component—which has a long refractory period and is not able to follow stimulation at frequencies of one per second without undergoing a reduction in height. This component is sensitive to the action of barbiturates and its appearance can be prevented by antidromic impulses to the bulb. This means that it is probably induced in the secondary neurons and is equivalent to the induced oscillatory waves.

Olfactory discrimination and electrical activity

The ability to distinguish one smell from another implies the recognition by receptors of certain distinguishing features belonging to odorous substances. This information has then to be transmitted to the brain in the coded form of electrical discharges. These processes are still almost entirely hypothetical and will be discussed as part of the 'Theories of Smell', but some evidence which appears to be of importance has been provided by recording the discharges in the bulb and these will be considered now.

1. SPATIAL PATTERNING IN THE BULB

Adrian (1953, 1956) showed that there are large differences in the spatial patterning of excitation produced by different substances.

He noted that if electrodes are inserted into mitral cells some will respond better to stimulus by certain odours and vice versa.

In the rabbit the threshold for water-soluble substances is lower in the anterior part of the bulb whereas that for lipid-soluble substances is lower in the posterior part. In the same manner acetone gives a discharge confined to the anterior part of the bulb and paraffin to the posterior part. Coal gas and its impurities sets up an intense discharge from the posterior part of the bulb without affecting the anterior. There is a possibility that the spatial pattern of excitation is due to differential distribution of the different odorous molecules in the nasal cavity, but there is evidence against this, namely the evenness of the olfactory mucosa in man and the fact that odours can be recognized if injected from behind the posterior choanae. If the excitation pattern in the bulb does reflect the distribution of molecules on the olfactory mucosa, then other factors such as the composition of the surface film and the velocity of the air current must also be important.

2. TEMPORAL PATTERNING IN THE BULB

There is also a difference in the time-course of the responses evoked by the different substances. Some produce a discharge of rapid onset and fast decay while others develop slowly and diminish gradually. Adrian found that the first, more abrupt, type of response was given by substances soluble to some extent in water, whereas heavier molecules, soluble only in lipoids, produced the more persistent type of discharge.

These responses are also to some extent similar to the receptor response, so that the temporal pattern of reception excitation is preserved at the level of the secondary neurons. The factors which determine this pattern are not known and there is no proof that they are indeed associated with discrimination mechanisms.

3. FREQUENCY COMPONENT ANALYSIS

Hughes and Hendrix (1967) carried out a different type of experiment on unanaesthetized conscious rabbits which had had stainless steel macro-electrodes implanted into the olfactory bulb.

With these they were able to record the rhythmical responses which were produced in the bulb as a result of stimulation with many different odours. These rhythmical discharges were then analysed into the various frequencies or components of which they were formed. The following features became apparent.

1. The lower frequency components of 4 to 5 per second represented respiration and sniffing rates and other groups under 20 per second were their harmonics, differences and summation frequencies.

2. There was an inverse relationship between frequency and amplitude of the major peaks and also between their frequency and the molecular weight of the odorant chemical.

3. The highest frequency components appear in bursts at the beginning of a discharge and do not vary with different substances.

These high frequency responses may thus signal the presence of an odour.

4. Lower frequency components, on the contrary, do vary in type with the different chemicals. It may be that these low frequencies signal the 'identity' of the odour.

5. There are both excitatory and inhibitory types of response.

6. Components of *high molecular weight*, especially floral odours often show their main response within the 37 to 75 per second range. Components of *low molecular weight*, especially camphor, respond between 75 and 125 per second.

7. The prominent frequency components appearing for a given category of stimuli vary according to the site of the electrode.

These findings, if confirmed, are of considerable importance as they show a general tendency to confirm a stereochemical approach to olfactory discrimination by the receptors; a similarity of response to the similar odours from frequency component point of view and provide further evidence for odour localization within the bulb.

Central regulative effects on the bulb

It would be no surprise to find evidence suggesting that the activity of certain brain centres can influence the afferent input from the sense organs. This has already been demonstrated in the muscle spindle, skin receptors and in the cochlea. In these organs the centrifugal regulation is exerted directly on the end organ but no such efferent fibres have been demonstrated passing from the olfactory bulb to the receptors. In the case of the olfactory system this centrifugal influence is exerted upon the transmission of the olfactory signals in the first relay station.

A system of fibres leading from the basal telencephalic areas to the olfactory bulb and from one bulb to the other had been known to exist since the beginning of the century. These last fibres are now thought probably to originate in the tufted cells. Stimulation of these areas of the brain causes a reduction of the spontaneous activity of the bulb and a reduction or even abolition of the bulbar response to olfactory stimulation. Similar effects were also obtained by high-frequency stimulation of the anterior commissure.

If the anterior commissure is cut there is an increase in the amplitude of the olfactory response. Weak olfactory stimulation in one nostril induces activity in the corresponding bulb, but when a strong odour is passed into the other nostril, the original induced activity will be depressed. These findings make it appear likely that this centrifugal system acts in the same manner as those controlling the input from other sense organs and therefore exerts a tonic influence.

The manner in which the excitability of these afferent olfactory relays is inhibited by fibres from the higher regions of the brain may be analogous to the influence of the recticular system on other sensory systems while the bulbs may be inhibiting each other through their connections.

Low frequency stimulation of the intra-laminar thalamus evokes in the bulb recruiting potentials similar to those recorded from the frontal cortex (Carreras et al., 1967). These bulbar recruiting potentials are superimposed on a sustained negative d.c. charge which increases as the frequency of stimulation is raised.

High frequency activation of the mid-brain reticular formation induces in the bulb a negative d.c. shift of 100 μV amplitude paralleled by a generalized arousal.

The significance of these findings is not clear but it appears that the thalamus as well as the reticular formation may influence the bulb. Auditory and gustatory stimuli have also been reported to produce activation of the olfactory bulb.

Depression and enhancement

It is difficult to speculate reasonably on this limited evidence, but in other sensory systems feed-back loops exist to depress the input of irrelevant information during focus of attention. It may be that the same requirements are present in the olfactory system. That some parts of the centrifugal system can enhance the efficiency of the olfactory mechanism also appears possible, and this state may be brought about by the brain stem during arousal.

Although theoretically higher centres cannot alter olfactory *discrimination*, since that is dependent on the nature of the exciting molecule, the fact that they exert what influence they have on the bulb may suggest some role on that function. It may be that the relative value of different odours to each other can be altered in that way, modifying the resulting smell or the individual's reaction to it. Could some types of olfactory distortions (parosmias) usually classified as 'functional' in origin be associated with a disturbance of this mechanism? Could, also, certain metabolic diseases such as adrenal malfunction mediate their olfactory changes through modifications of this type rather than through the raising or lowering of the threshold?

THE CENTRAL OLFACTORY CONNECTIONS

For many centuries philosophers and scientists searched for the seat of the soul and many viscera were said, in turn, to contain it or at least to house certain aspects of its emotions. The cerebral cortex appears never to have been considered, and it is only recently that its function has received the objective attention which a study of its electrical discharges can give.

Investigation of the olfactory part of the sensory system has been complicated in man by the fact that the great development of the neopallium has pushed the structures of the rhinencephalon or limbic system onto the medial and under surface of the cerebrum. In this position they are almost inaccessible to experiment and to surgery and the electrical discharges which are generated there are far from the areas detected by the more common electrodes.

The use of depth electrodes inserted into specific areas of the brain by stereotactic apparatus has produced information of great value, particularly when they can be used both to record and to stimulate.

In 1942 Adrian showed that olfactory stimulation gave rise to a train of regular waves in the *pyriform lobe*. Other workers have shown that electrical stimulation of the olfactory bulb produces responses in the *prepyriform cortex*, the *anterior olfactory lobe*, and the *olfactory tubercle*.

Sem-Jacobsen and his colleagues (1953) inserted two electrodes in the region of the left cribriform plate and recorded a rhythmic discharge when various odours were sniffed up the left nostril but not the right.

Walter (1958) stimulated electrodes inserted into the base of the caudate nucleus and the anterior perforated substance of a conscious patient who then reported the production of a sensation of smell.

Stimulation of the amygdala

The function of the complex of grey matter which is placed deeply into the substance of the temporal lobe and which is called the amygdala has evoked many controversies. Many experimenters have studied its place in olfactory function, but one of the earliest was Allen (1941) who found impaired olfactory discrimination in conditioning experiments after bilateral amygdalectomy.

Earlier still, Johnston (1923) had suggested the hypothesis that the amygdala may be a complex in which olfactory, gustatory and general somatic sense impressions are brought into correlation.

Creutzfeldt, Bell and Adey (1963) tested this hypothesis by recording the reactions of single neurons in well localized areas of the complex using anaesthetized cats as experimental animals. Electrical stimulation of the olfactory bulb and stimulation of the mucosa by cigarette smoke were used as well as acoustic, visual and somato-sensory stimuli. The responses were divided into 'specific' and 'diffuse' types depending on whether they were stimulus-bound and showed a fixed reaction pattern, or whether they were inconsistent with an ill-defined pattern. The diffuse type also had a longer and more variable latency. The olfactory input area was found to be almost entirely in the *mediobasal* and *cortical nuclei* and the *anterior area*. Real multisensory convergence was found only in the central nuclei and the globus pallidus.

Chapman (1960) reported some interesting studies made in unanaesthetized patients. Most of these had temporal lobe epilepsy and depth electrodes had been inserted into the amygdala using stereotactic apparatus so that they could be electrically stimulated and later coagulated. Two cases are worth noting. One, an 18-year-old female was mentally deficient and showed intractable assaultive behaviour. Electrical stimulation caused her to look startled and to smell ether. She would repeatedly hold her nose during stimulation and cry out: 'Get that stinking stuff out of here' or 'There is that stuff again'. The other was a 39-year-old male who had experienced seizures of grand mal type as a result of a kick in the forehead by a horse during childhood. Electrical stimulation of the amygdala reproduced an olfactory aura.

The pyriform lobe
The electrical discharge obtained from the surface of the pyriform lobe during olfactory stimulation shows a negative wave preceded by a spike potential. This is contrary to the response of other sensory areas which show, instead, a positive wave. The reason for this may be that the olfactory tract fibres approach the olfactory cortex from its surface while other sensory systems are relayed from deep thalamic centres.

These waves can be shown to be composed of two negative waves, one which is augmented by repetitive stimulation and the second which is blocked. It may be that this last wave originates

c

in the transcortical connections between the pre-pyriform and pyriform cortices.

The distribution of the negative surface potentials evoked by stimulating the olfactory bulb closely corresponds to the cortical distribution of the olfactory tract fibres. Around this area, which is the primary olfactory cortex, a positive potential is produced and this represents the activity of areas analogous to secondary areas in the other sensory systems.

The role of the hippocampus

Anatomical and electrophysiological investigations have shown that the olfactory fibres do not supply the hippocampus. Together with the fornix, and cingulate gyrus, it may have lost its olfactory function, if it ever had any. Deeper parts can be activated by visual, tactile, auditory and olfactory stimuli only after long latent periods indicating several synapses and with no evidence of localization. It is probably a true associational region, where vision, hearing, and somatic senses may each be effective even on the same cell.

Walker, Andy and Faeth (1954) have speculated on the basis of anatomical connections, that the proprioceptors, interoceptors and exteroceptors are integrated through the thalamic mechanisms; but the teleoceptors, audition, vision and olfaction may be brought together mainly at a cortical level in the temporal and occipital lobes. From there they are projected to the opposite hemisphere and sub-cortical structures by means of the amygdala and hippocampus. These then, are the ganglia of an effector system operating under the influence of distant receptors. The effector mechanism of these structures would influence not only cortical, but sub-cortical motor and autonomic centres in order to initiate direct responses to change in the distant environment. In the case of smell the actual cortical representation remains in the pyriform and prepyriform areas.

CHAPTER FOUR

Theories of Smell

The first steps in the development of theories of smell must be the attempts at classification of different smells. The early methods were based on the assumption that odours could be grouped according to similarities in their perceived qualities and the procedures used were therefore entirely subjective. Among the innumerable classifications those which have attracted most attention have been Rimmel's, Zwaardemaker's, Henning's and Crocker and Henderson's.

Rimmel's, which is the oldest of them, divides the pleasant odours into 18 classes of which each contains similar smelling odours. Thus with 'fruity' is included pear, apple, pineapple and quince, while 'spicy' would include cinnamon, cassia, nutmeg, mace and pimento. The 18 groups were the following:

1. Rose	7. Spice	13. Mint
2. Jasmine	8. Clove	14. Aniseed
3. Orange Flower	9. Camphor	15. Almond
4. Tuberose	10. Sandal	16. Musk
5. Violet	11. Citrine	17. Ambergris
6. Balsamic	12. Lavender	18. Fruit

Zwaardemaker took the problem much further in 1895. He not only produced nine associative classes, but suggested that there must be nine different types of cells in the olfactory epithelium. This was clearly the beginning of a Theory of Smell and he tried to substantiate it further by fatigue and adaptation experiments. Zwaardemaker's nine classes are the following:

1. Ethereal
2. Aromatic
3. Balsamic (fragrant)
4. Ambrosial
5. Alliaceous
6. Empyreumatic

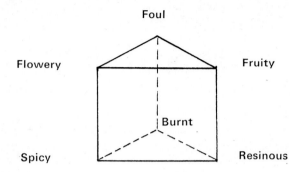

Figure 4.1   Hening's prism.

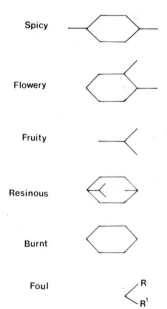

Figure 4.2
Hening's chemical classes of
odour.

7. Caprilic
8. Repulsive
9. Nauseating (or fetid)

In 1916 Henning produced a new classification by reducing the number of classes to six, but suggesting that they should not be separate but that there were intermediate odours merging the classes into one another. He illustrated his arrangement by an Olfactory Prism (Fig. 4.1). Thus fundamental odours would be placed at a corner and intermediate odours would be placed along the intercommunicating sides. He developed his classification into a theory by showing that certain chemical groups placed in particular positions to the rest of the molecule could be related with the six classes, which are (Fig. 4.2):

1. Spicy substances were associated with the para-arrangement in the benzene ring.
2. Flowery with the ortho-arrangement.
3. Fruity.
4. Resinous. The groups are within the ring.
5. Burnt. A smooth ring.
6. Foul substances have a sulphuretted arrangement.

In 1927 Crocker and Henderson postulated four different kinds of olfactory nerves, each responding to one of the following four groups of odours:

1. Fragrant
2. Acid
3. Burnt
4. Caprylic

Each of these fundamental odours was given an intensity number of 0 to 8, then a complex odour could be represented by a row of 4 numbers each representing one of its fundamental odours by its intensity.

All these attempts at classification suggest a manner in which the physicochemical properties of the odorants are associated with their smell and in turn this has led to Theories of Smell. We now know that smell is a complex process starting with the stimulation of the olfactory epithelium by air bearing odoriferous molecules. This sets up a transmitted electrophysiological impulse along the olfactory pathways of the brain until ultimate perception takes place. At present the nature of each of these processes is still a matter for speculation and research and it is therefore worthwhile considering the views which have been expressed in the past.

Since early times theories based on attempts to correlate the chemical and physical properties of an odorant to their smell have multiplied into an increasing degree of confusion.

The theories of smell can be grouped into two main divisions:

1. The corpuscular theory

The suggestion here is that the olfactory region in the nose is stimulated by odorants which are carried in the air in particulate form. Some sort of chemical reaction takes place on contact with the sensory surface.

2. The wave theory

In this theory it is suggested that by analogy with light and sound, odorant substances emit waves which then stimulate the olfactory organ.

There are many variations and modifications of both these theories including a corpuscular theory which offers vibration as a mode of stimulation of the receptor surface by the particle. Many theories postulate volatility and solubility of the odorants as without an appreciable vapour pressure the molecules cannot make contact with the olfactory epithelium and they must equally be soluble in its lipoid matter. The importance of the intramolecular vibrations has also been postulated.

On the whole, the corpuscular theory has been the most favoured, but a revival of wave-theory was produced by Beck and Miles (1947). Their suggestion was that the olfactory receptors of animals were sources of infra-red radiation which radiate at different wavelengths according to their size and shape. Odorant substances have absorption bands in this region and if they come into the field of radiation of the receptor region, they will absorb the appropriate infra-red radiation. This will stimulate the radiation energy of that particular wavelength from some of the receptors. This loss of energy becomes the initial event in the process of stimulation of the olfactory receptors. According to Beck and Miles, the critical region in the infra-red spectrum is around the maximum of a black-body at 290 to $313°K$ (17 to $40°C$). Quality of odours is accounted for by differing spatial stimulation of the different receptors. The experimental evidence given in favour of this theory is based entirely on insects and there has not been general acceptance.

Indeed, there are more objections to it than there is supporting evidence, even circumstantial. Although butyl alcohol and its deuteroxyl counterpart have different infra-red spectra they have exactly the same odour and again, ordinary and heavy water also have different spectra yet neither has a smell at all. On the other hand the d- and 1- forms of some optically active compounds have different smells although their infra-red spectra are identical.

Of the theories which demand the arrival of a particle of odorant substance at the olfactory epithelium, again there have been many suggestions as to exactly how the receptor organ is stimulated.

1. Chemical theories

These include the 'osmophone' groups of Henning as well as the essential water and lipoid solubility. One important observation was that first made by Woker in 1906, that saturated compounds rarely produce odour and then only if unusually volatile. Unsaturated substances were more odorous and indeed, the odour increases with unsaturation. This was an important discovery and was elaborated by Durrans (1920). He noted that it was common for odorous substances easily to form addition compounds—in other words that they had a free residual affinity. He called this the Residual Affinity Odour Theory and suggested a probable course of events. The smell molecules on reaching the olfactory mucosa dissolve in the aqueous mucus layer which bathes it and then a part, the size of which depends on the oil/water partition coefficient of the substance, passes into the lipoid layer. In this layer an addition reaction takes place with the odorous molecule and the change of energy which results stimulates the olfactory nerves. The quality of the odour would then depend on the nature of these addition reactions and the rate at which they take place. The big drawback to this theory is the fact that saturated paraffins, which should be odourless, can in fact have a strong smell if sufficiently volatile.

Other chemical theories involved chemical reactions between the smell molecule, the substance chemicals of the olfactory cell or nerve ending, oxidizability of the odour substance, volatility, solubility and the oil/water partition coefficient.

2. Physical theories

These are not as numerous and the most important is Müller's Dipole Theory. The variety in smell is due to differences in the dipole moment. A molecule has poles if the electric centre of its positive charges does not coincide with the electric centre of its negative charges. In order that the moment can be made more noticeable the molecules have to be separated from each other so that they produce little electrical interference. The polar molecules irritate electrically the molecular field of the smell-receptor substance and this is transmitted to the olfactory nerves. Non-polar molecules such as benzene, hexane, carbon-tetrachloride, have similar smell, while substances such as indole and farnesol produce their smells when highly diluted. The smell of the non-polar bodies is an important argument against this theory but there does seem to be an association between polarity, chemical reactivity and the odour of substances.

3. Electro-chemical theory

McCord and Witheridge (1949) observed that in all molecules of three or more atoms the electron arrangement is understood to form bonding angles between the norms. The potential maximum number of bonds is one less than the number of atoms forming the molecule. They speculated that when an odorous substance dissolves in the olfactory mucosa, changes takes place in the bonding angles and the electrostatic fields are altered. The potential difference so produced stimulates the receptors. The most important defect of this theory is the odour of diatomic molecules such as chlorine. According to this theory they should be odourless.

4. The Stereochemical theory

Lucretius is generally credited with being the first to record a theory in which the shape of the molecules is related to the quality of the smell. Writing two thousand years ago he could hardly be considered as a precursor of the understanding of molecular structure, but he can be placed firmly in the stereochemical camp.

There is no doubt that the first person to state clearly this promising idea was Moncrieff (1949). He had previously (1944) rather tentatively put forward a theory that the only two pre-requisites for a substance to be odorous were volatility and a suitable solubility, even if this were to be a rather special type of solubility such as that in protein. The close link between molecular morphology and solubility has long been known so it was only a step to postulate that the different shapes of molecules will fit on to certain available molecular sites in the olfactory receptors. This type of process is quite often found in therapeutics and pharmacology and the specific solubility of the substance is only one expression of the molecular shape. A shape theory explains why

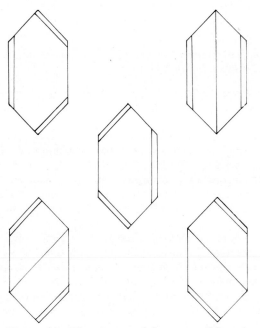

Figure 4.3   Five forms of benzene. Although this represents how a molecule can find different forms, modern chemical theory cannot, of course, accept such a simple concept in the case of benzene.

unsaturation or residual affinity is often associated with powerfully odorant properties, as unsaturated bodies often have a structure which permits resonance, the molecules constantly changing their shape. Thus bezene may exist in five forms:

The greater the number of shapes a molecule can take, the greater the   possibility of fitting into one of the available sites. Double bonds are points of weakness and can be distorted but if the double bond is close to a polar group it will be still more unstable and easily distorted. This plasticity will help it to fit into a receptor site and it is certainly the case that the molecules are almost invariably odorous. Thus it is the plasticity of the unsaturated compound which will give it its smell and not its additive reactions.

Moncrief's theory of odour demands then two pre-requisites from odorous substances:

1. Volatility.
2. A molecular configuration that is complementary to certain sites on the receptor system.

This theory has the great advantage not only of explaining the phenomena but also of embracing the valid features of previous theories.

The second important contribution to the stereochemical theory and to the study of odour was provided by Amoore in many papers from 1952 onwards. He felt that to make practical use of this speculative theory two problems had to be solved. Firstly to identify the primary odours of the sense of smell and secondly to discover the physical dimensions of the corresponding receptor sites.

Amoore began by making a list of pure compounds of known chemical structure all possessing the same smell. This was done in an entirely theoretical manner by surveying the chemical literature. Having done so, he looked at all the characteristics of these molecules in an attempt to discover the common ones. The features which Amoore found important seemed to vary from smell to smell; in some cases it was the size of the molecule, in others the shape, and in yet others it was the electronic status which was common. To help in this computation molecular models were used and as the most valuable information would be provided by the more rigid molecules he divided them up into three types based on the degree of certainty of a single spatial configuration. He called them 'invariant', 'determinate' and 'articulate'.

Amoore compiled in this way a list of more than 600 odour descriptions from the literature. This list could be divided into 14 sub-lists according to their different odours:
Camphoraceous; Pungent; Floral; Ethereal; Minty; Musky; Putrid; Almond; Aromatic; Aniseed; Lemon Cedar; Garlic; Rancid. The chemicals in each list were extremely diverse from all

chemical aspects and Amoore tried to identify the primary odours according to the stereochemical site-fitting concept.

From the principle of probability and by imagining receptor-sites of different shapes, rigid molecules able to fit a given combination of two or more sites are likely to be far less common than those able to fit one size only. In other words, if one took the lists of odorants, certain odours should be rather more common and these are likely to be the primary ones. Those which are rare are likely to be multiple-primary or complex odours.

Thus, the primary odours would appear to be:

> Camphoraceous
> Pungent
> Floral
> Ethereal
> Minty
> Musky
> Putrid,

while

> Almond
> Aromatic
> Aniseed

smells appear to be border-line cases.

Scale models of these molecules can be constructed from knowledge of similar compounds, X-ray crystallography, electron diffraction, and other methods. These facts have gradually been recorded and are now available in text-books.

If it were accepted that receptor-sites for these molecules really exist, then their shape can be worked out, as we know the dimensions of the molecules which must fit into them. This has been done for ethereal, camphoraceous, floral, musky and minty smells and the shape of the receptor-sites drawn to scale (Fig. 4.4).

Pungent-smelling compounds are almost all highly reactive chemically. The radicals which are present render the molecule strongly electrophilic so that it would not be surprising that the receptor site contains nucleophilic groupings such as a sulphydryl SH group. In this way the odour molecule may form addition compounds with the site molecules or oxidize or alkylate them. The interesting feature here is that pungency also stimulates common sensation mediated by the trigeminal nerve, as all strong chemicals produce common as well as olfactory sensations.

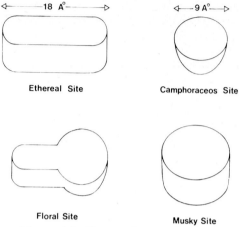

Figure 4.4 Shapes of some olfactory receptor sites.

Putrid-smelling compounds have, on the contrary, strongly nucleophilic atoms. These are often trivalent nitrogen, phosphorus, arsenic, divalent sulphur, selenium and tellurium. A corresponding electrophilic site may therefore well include a heavy metal ion with which the odorant molecule might react or form a co-ordination compound.

When Amoore tried to elucidate the sites of the less common odours these became less distinct and the molecules of the odorant could fit into two more primary sites. He suggested, therefore, that complex odours were caused by concurrent stimulation of two or more primary sites. If this were the case, it should be possible to reproduce any complex odours from the appropriate primaries by mixing them in the right proportions. This has been done using theoretical stereochemical principles in the case of cedarwood (Jonston, 1963) from camphor, musky, floral and minty primaries.

5. The vibration theory

Although waves, of course, involve vibration, it should be made clear that when we now refer to vibration theory it is understood that odorant particles actually come into contact with the olfactory area. Intramolecular vibration is involved in stimulating the organ and not waves set in motion by a distant source. In this way the vibration theory is really a special case of the corpuscular theory.

Vibration theories are not new and in the early years of this century when electron theory was attracting a great deal of attention, Teudt expounded his version in the German-language journals. His most important contribution was that odour was a property of molecules and not of atoms. Teudt believed that it was the electrons shared between the atoms of a molecule which themselves stimulated the olfactory nerves. The valency electrons have a rhythmic motion as they are alternately attracted and repelled by the atoms which they share. The non-valency electrons do not participate in this motion and the period of vibration depends therefore on the nature of the molecule and varies from one to the other. The molecules of the olfactory nerves also vibrate and when these correspond with the period of an odorous substance, resonance occurs and a nerve stimulus is produced in the same manner as an induced current can be obtained. Odorous molecules can induce vibrations in non-odorous molecules which can then take on different odours. Although the molecule does not have to come into actual contact with the olfactory nerve, it would nevertheless have to be very close. Although this was a very interesting theory and the idea of intramolecular vibrations could explain many of the features of smell, there is nothing to confirm the suggestion of induced resonance which must remain in the realm of the fantastic.

A decade or so later Krisch elaborated a general theory of the senses based on the vibration hypothesis. He made two general assumptions: first that all energy is basically of the same nature and, secondly that in the development of early animals there was originally a single sense organ which 'perceived' the energy produced by external stimuli. As he believed that the present sense organs were all evolved from this original single one, it was reasonable to expect the nature of the response to stimuli to be similar. Thus light, sound, smell, etc., all depend on electronic vibrations of different magnitudes to stimulate sense organs modified to respond to these magnitudes. In this way there should be an olfactory spectrum as there is a visual one.

Heyninx, a contemporary of Teudt and Drisch, also supported a vibratory theory, but this time stressed that actual direct contact was necessary for odorous molecules to stimulate the olfactory area. He said that the wavelengths of these odorous and ultra-violet vibratory motions lay between 0.35 μm and 0.20 μm and that they grouped themselves into seven fundamental groups: Ethereal,

Vanilla, Spicy, Burnt, Offensive, Rotten and Sharp. Adsorption took place on to the olfactory epithelium where the odorous molecules formed a layer 0·13 μm thick. The olfactory spectrum vibrated within a range of 0·35 to 0·20 μm in wavelength and this caused vibration of the hairs of the receptor cells, the whole being intensified by the resonance of the olfactory pigment.

Ungerer and Stoddard (1922) from America, produced another vibrational theory. They accepted the fact that an odorous molecule must come into contact with the receptor surface but suggested that the stimulus resulted from the intra-molecular vibrations of the substance and not from any chemical or physical reaction. Every pure substance had its own rate of vibration forming a continuous spectrum and those which fell within a specific range were perceived as smell. These vibrations from different molecules can combine to produce other frequencies. This would explain the strengthening and 'body' added to a feeble perfume by a small quantity of certain other substances (see Chapter 11). In the same way such combinations can explain the disappearance of odours on purification. Despite its interesting qualities there is considerable difficulty in explaining certain olfactory phenomena such as fatigue within the context of this theory.

At the present time Wright is the main protagonist of vibrational theory and he traces his concept to the work which Dyson produced during the 1930s.

Dyson's theory was based on the hypothesis that it is not the size, shape or reactivity of the molecule which forms the physical basis of odour but its vibrations. What places Dyson's work in a different category from all the previous theories of vibration is his introduction of tests for his hypothesis instead of resting on speculation. This was possible because of the development of the Quantum Theory and the discovery of the Raman Effect and its observation.

Atoms consist of both positive and negative electric charges so that when they or the molecules they form oscillate they can produce and interact with electromagnetic radiations. If infra-red light is passed through such molecules some of the wavelengths which comprise it will be the same as those of the intramolecular vibration. These will resonate and be absorbed, so that when the light is later passed through a prism a spectrum will be formed with dark absorption bands whose frequencies correspond to the vibrational frequencies of the molecules. In this way the vibrational frequencies of odorous molecules can be demonstrated.

Dyson suggested that the Raman Effect which at that time had only recently been discovered would be the best way of testing his hypothesis. This Effect is based on an understanding of Quantum Theory which had become increasingly widespread since the work of Max Planck at the beginning of this century. When an atomic or molecular particle or even an electric particle is given an oscillating motion it will have certain quantites of energy of motion or quanta. There are never any fractions of quanta and the size of a quantum depends on the frequency of this oscillation. The frequency is itself dependent on the force and mass; if E is the energy of the oscillator, n is the quantum, v the frequency and h is Planck's Constant, therefore:

$$E = nhv$$

If monochromatic light (i.e. an electro-magnetic radiation with a single frequency) is passed through a substance, some molecules would begin to vibrate. Given the frequency of the incident light is vi then its energy is hvi. If some molecules have begun to vibrate as a result they must have taken the energy from the light whose energy and therefore frequency must be reduced, say, to v_o then the energy removed is:

$$hv_i - hv_o$$

This energy in turn must be the same as that of the recipient vibrating molecule, hv, or:

$$hv = hv_i - hv_o$$

therefore

$$v = v_i - v_o$$

In other words this effect, the Raman Effect, means that by measuring the *difference in frequency* of the light before and after its scatteringf by the molecules of a substance, it will give us the vibrational frequency of the molecule. This technique, which involves light, works with transparent substances.

Dyson pointed out that a vibrational theory could explain how musks which had different chemical structures could smell alike and how ketones which are so similar in structure could smell so different. He suggested that the 'osmic frequencies' which stimulate the olfactory apparatus of the nose lay between 1400 and 3500 wave numbers. This 'wave number' is not really the frequency (oscillations per second) but is the number of waves per centimeter and in order to obtain the frequency it has to be multiplied by the speed of light.

The great advantage of Dyson's theory was that it could be tested using infra-red and Raman techniques, but when it came to proof the frequency range of 1400 to 3500 wave numbers turned out to be incorrect. Odours could not be predicted from the frequencies so this brave attempt at a theory which could be tested instead of one which was based on speculation had to be abandoned.

In 1954, however, Wright resurrected the theory and made it more specific, giving reasons for supposing that the relevant vibrations are of low frequency and likely to have a 'whole molecule' character. He noted that in order to vibrate, a molecule must have at least one quantum of energy, but the problem was how it could acquire this energy in the dark confines of the nose. The common energy sources such as heat, chemical reaction or light are, of course, not present, and Wright suggested that the energy to excite the vibrations of the odorous molecules must be the collisions they make with the nitrogen and oxygen molecules of the air.

The temperature inside the nose is only 30°C to 35°C which is about 300 K in absolute terms. As the violence of inter-molecular collisions is directly proportional to the absolute temperature the relatively small amount of energy made available in this case sets only low-energy vibrations going. Wright calculated the average number of quanta of vibrational energy that could be given to vibrators of various frequencies by collisions with air molecules at 300 K. This showed that it is only molecules with a low frequency, below 200 wave numbers, which have an average one quantum of vibrational energy. Higher wave numbers produced many more 'silent' molecules than active ones.

Wright could show that 'osmic frequencies' had to be below 500 wave numbers to function in the conditions of the nose, so that Dyson's frequencies of 1400 to 3500 wave numbers were totally impossible.

Measuring the low-frequency vibrations of almond-smelling compounds, Wright produced the results presented in Table 1.

TABLE 1. *Low frequency vibrations of almond-smelling compounds*

Nitrobenzene	Benzonitrile	Nitrothiophene	Butyronitrile	Benzaldehyde
176	172	169	179	130
252				225
	320			237
397	381	376	370	439
435	405	442		
532	460		524	
	549			

This shows some correlation but is far from convincing by itself. There are also many difficulties in testing the theory because of the problems of Raman or infra-red spectroscopy. All the lines observed do not represent fundamental modes as there can be overtones and combination bands of two different fundamentals. Examination of the common features in the far infra-red spectra of groups of compounds with similar odours has shown a few statistically significant correlations in certain patterns of vibratory frequency and some musks, bitter almonds and cumin (Wright, 1967; Wright and Robson, 1969). In an experimental test of this theory Wright and Michels (1964) chose nine compounds as odour standards and 50 unknown. Their similarity was rated 0 to 6 by a panel of 84 judges and the similarities in the Raman spectra assessed in terms of the sum of the spectral intensities occurring within particular wave-number intervals. The table of odour similarity ratings was converted into a 50 by 50 correlation matrix which was factor analysed by computer, yielding eight factors. When the odour factor loadings were compared with the Raman spectra significant correlations were found ($P \sim 0.01$). But even this probability level falls far short of that achieved by the stereo-chemical theory in a comparable survey with factor analysis ($P < 10^{-9}$).

6. 'Penetration and puncturing' theory

Discussion of the previous theories has suggested many problems which have to be explained in order to make the theory acceptable. It should conform with known physiology, it should explain the mechanism by which the nose discriminates between the innumerable odours, particularly those which are only slightly different. It should also account for the extreme sensitivity of the human nose, not to speak of animals. Man can perceive as little as 10^8 molecules of β-ionone. But perhaps the key problem is the exact manner in which the stimulating molecule alters the cell and so generates an impulse. It is likely, from the behaviour of other physiological mechanisms, that in some way the permeability of the cell wall is altered, but the manner in which this comes about remains to be shown.

The cell membrane is generally taken to consist of a bimolecular layer of lipids with a protein layer on either side. This can be represented in cross-section as in (Fig. 4.5). The protein layer, which is drawn as a single thickness, consists of branching molecules, so that if imagined in three dimensions it will form a mesh with openings in it. The lipid molecules consist of an outer polar

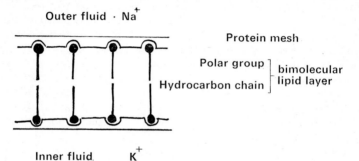

Figure 4.5   Representation of cell-membrane.

group and a hydrocarbon chain pointing inwards. There is also a layer of long, branching polysaccharide chains standing normal to the outer protein layer. The fluid on the outside of the cell has a high sodium content while that on the inside a high potassium concentration, and as far back as 1949 Hodgkin and Katz were able to show that when the axon was excited the potassium ions were able to escape through the cell membrane. At the same time sodium ions enter the cell tending to normalize the ionic imbalance. When excitation has passed and the cell reverts to its resting state sodium ions are pumped out through the cell wall and potassium ions re-enter. As this goes against the gradient it is referred to as 'active transport'.

Takagi, Wyse and Yafima (1966) showed that the depolarization of the olfactory membrane depended mainly on an influx of chloride ions and only secondarily on the movement of potassium ions. Studies showed that F⁻, Cl⁻, Br⁻ and HCO_2^- can penetrate the olfactory membrane while other anions such as I⁻ and ClO_3^- cannot. Those that cannot are the larger ions so this suggests that there are punctures in the membrane acting as sieves. Although thin, the bimolecular lipid layer of the cell-membrane is enough to render the cell impermeable and as the hydrocarbon chains can move about it has a certain liquid quality which allows it to preserve its continuity.

In 1953 Davis produced a theory of smell which supposed that the large, rigid, often awkwardly shaped odorant molecule would penetrate the cell-membrane. If it diffused through it would leave behind a hole in the membrane which would heal relatively slowly so that ions can leak through. This exchange of ions could then initiate excitation of the nerve by short-circuiting the membrane resting potential. The generator current which would result does not itself travel along the nerve fibre, but it triggers off the nerve impulses which probably originate where the receptor cell narrows down to the olfactory fibre (Fig. 4.6).

Large molecules such as β-ionone are most effective because a single one can cause significant leakage, whereas in the case of smaller molecules such as ether or chlorophenol several have to penetrate side by side in order to do so and would be less sensitive.

To place their theory on a quantitative basis, Davis and Taylor (1957) and Davis (1962) assumed that two factors control the dislocation of the membrane:

1. The adsorption coefficient (adsorption energy) of the smell molecules in passing from the air the lipid-water interface consis-

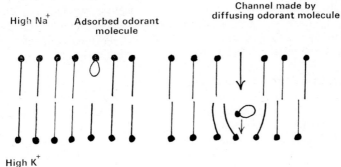

Figure 4.6   *Diffusion of adsorbed odorant molecule (after Davies).*

ting of the lipid wall and the layer of mucus which bathes it. This can be calculated from the free energies of the molecules in passing from air to water and then water to lipid.

2. The capacity of each adsorbed molecule to puncture the membrane. Large molecules can do this singly while smaller ones require a number p (5 to 10) to be absorbed side by side. In that way Davis and Taylor (1957) were able to derive an equation correlating olfactory thresholds in terms of molecular size and adsorption energies of odorant molecules. As the adsorption energies can be obtained from experiments and the size of the molecules is known, thresholds should be predicted.

The quality of the odour will depend on the rate of diffusion of the odorant through the membrane relative to the subsequent rate of healing, so the intensity of the odour depends on:

$$1 - \frac{t_{\mathrm{diff}}}{t_{\mathrm{h}}}$$

if t_{diff} is the time for diffusion through and t_{h} is the time of healing of a membrane. If t_{diff} is small compared to t_{h} a sharp hole will develop. With a large molecule the diffusion time may be great compared with healing time, so that the hole sharpness would be reduced. If,

$$t_{\mathrm{diff}} = t_{\mathrm{h}}$$

then the penetrating molecule would not leave a hole and no ions would escape.

Furthermore, the membranes of the different cells may well vary in their healing capacity, thus producing differing t_{h}. Thus musks would stimulate only cells with slow-healing membranes, while small molecules such as ether which are too weakly adsorbed would not, and vice versa. Davis (1969) developed a quantitative theory for musks which could be used experimentally.

Discussion of theories of smell

It is clear from the preceding description that no clear-cut dominant and accepted theory has yet emerged. A month does not pass before a new variant is proposed, but on the whole some experimental evidence is usually given, which is an improvement on the past, where speculation differed only in the degree of cleverness.

To anyone attempting to follow this active field the picture becomes confusing rather than clearer, particularly when statistical techniques are given great importance. In an area such as this where chemical experiments necessarily carry some error and where psychophysical methods are notoriously aberrant, it is easy to be misled by masses of figures, matrices, factorial analyses, etc. The danger lies where discussion is limited to the statistical validity of the result and the experimental error which is inherent in these forgotten results.

Nevertheless, an attempt at clarification is necessary and the first question that has to be asked is: 'When we are still not in possession of all the facts, do we need to formulate a theory at all'?

The answer depends on the motives of the person involved. Where academic curiosity is the reason or an attempt at advancing knowledge of normal physiology, it could be no. Perhaps some theory will suggest further experiments, but to a large extent the mind should be kept as open as possible.

On the other hand, the majority of the workers in this field are scientists connected with the perfumery or the food processing industries and their need is for a working or workable theory— one which permits them to plan and conduct their work in a progressive manner even if it may be proved wrong or incomplete later. In other words, there should be two considerations when considering a theory. The first is an ideal one—its 'absolute correctness'. The other is a sort of second best when only some of the facts are known. In that case the chosen theory should explain the known facts but it should also be useful in a practical sense, permitting at the same time the next step forwards.

If a theory is valuable to practical workers it is no use criticizing it according to the first, ideal, consideration. It should be accepted for what it is, ready for modification and reconsideration.

The author of this work is a doctor in clinical practice and when faced with a situation where many theories abound, none proved beyond doubt, he must decide which theiry will be most useful clinically. The hope of course remains that the results of such clinical investigations as may be possible could be a small contribution towards the final outcome.

There is no doubt that to an ear, nose and throat surgeon the most immediately acceptable theory is a vibrational one. We are already familiar with methods of measuring hearing thresholds for different pitches and referring to them as so many 'cycles per second', and nothing could be more attractive than to transfer all this to smell. If the vibrational theory had provided us with a spectrum of vibrations, each responsible for a different smell, the direction of smell-assessment would be clearly defined once and for all. This, however, is not the case, although statistically significant correlations have been established between certain vibratory frequency patterns and the odours of musk, bitter almond and of cumin. How the vibrational features of an odorous molecule selectively stimulate the receptor cell to express its specificity is not yet clear and it cannot at this stage be applied to or tested by the clinical situation.

The puncture theory is even less useful in that context. Furthermore, it is difficult to accept the idea of a molecule penetrating the cell through the lipid bilayer and then disappearing, leaving a hole which does not heal immediately. One would imagine that a surface of that type shows no specificity to isomers as they have a similar structure; and yet many optical isomers have entirely different smells. There are a number of variants of the puncture theory which suggest that adsorption on to the surface may be enough to modify it sufficiently to alter the ionic balance. The whole structure of the cell membrane is itself in doubt. Many have suggested that apart from the protein and lipid layers there is an external 'fuzz' consisting of mucopolysaccharide and mucoprotein. This layer is a collection of highly hydrated networks whose volume can be altered by 100,000 times by minor changes in the calcium concentration, and it is most unlikely that it is without a role in the generation of cell excitation. The protein layer itself is probably vital in the formation of the membrane potential, and, according to Eigen and de Maeyer's recent work (1969), solium and potassium ions do not move alone but are carried by peptide carriers. It is difficult to see how the puncture theory can help us clinically at the present time.

Amoore's work has appeared to the author to be most easily applicable. The idea of shapes fitting into receptor sites is now quite commonly expressed in pharmacology and therapeutics. The most useful part of his work has been his early classification of substances according to odour and his discovery that seven odours are very much more common than the rest. Without necessarily accepting that these are 'primary' and the others 'complex' it seems entirely reasonable to use these common odours as test material in qualitative investigations, thus giving these some point of contact with the different type of approach used by the chemists.

CHAPTER FIVE

Testing the Sense of Smell — Olfactometry

The great advances that have been made in the study of hearing and vision were entirely dependent on the accurate measurements of these perceptual phenomena. Furthermore, some measuring instruments of great precision have been devised to study these senses both subjectively and objectively.

Nothing of this calibre has appeared in olfactometry, and this lack of adequate measuring instruments has been particularly unfortunate because of the rather limited knowledge that we have of the fundamental manner in which this sense functions. Wenzel and Sieck (1966) have complained that every experimenter or clinician has insisted on designing his own olfactometer and that there has therefore been no consistency or indeed comparison possible between the different reports. They also express astonishment at the simplicity of some of the techniques offered—techniques which make no attempt at controlling important parameters such as temperature; humidity; purity of the background air; flow rate, duration and concentration of the odorous air.

There is no doubt that these criticisms are important and, because they cannot properly be ascribed to the incompetence of olfactometrists as compared with other physiologists and clinicians, they must be central to the whole problem of the measurement of smell. In this chapter the techniques used in these attempts will be reviewed and those which appear most representative or most relevant will be described in detail. At the end, however, Wenzel and Sieck's two critical points must be considered.

In the past olfactometry was entirely subjective, but more recently some attempts at introducing objective methods have been made. Nevertheless, we can still divide the tests into two broad groups: subjective and objective, leaving the rather exceptional haematogenous methods as a separate small group on their own. The tests of adaptation and psychophysical techniques are best discussed on their own.

Among the welter of tests the experimenter will also have to choose and modify according to his purpose, whether it is a physiological experiment or a clinical test.

I. SUBJECTIVE TESTS OF SMELL

A large number of different approaches have been tried and many systems devised to classify them. They will be considered under the heading of Minimum Perceptible Odour measurement using simple and complex tests, test of recognition and certain special tests. The investigation of fatiguability and adaptation involves subjective responses as do the more recent psychophysical methods but because of their exceptional aims these will be discussed at the end of the chapter.

1. Measuring the minimum perceptible odour by simple tests

(a) Dilution in air

This must have been one of the earliest methods, as Valentin described a technique for diluting smells as far back as 1848.

He sealed a known quantity of an odorous substance into a fragile tube of very thin glass. This he placed into a bottle of known volume which was then closed. The thin tube can be broken by shaking the bottle and the person being tested can sniff at its neck. The experiment can be repeated at greater dilutions by using larger bottles for the same amount of odorant until a threshold is reached. This measure would then be in milligrams per millilitre of air.

(b) Dilution in an inert substance

Odorous substances can be mixed with an inert powder so that the intensity of the smells is the same. The threshold of smell as a minimum perceptible odour can then be measured on a linear scale in centimetres by gradually increasing the distance between the subject with his eyes closed and the mixture. Froelich (1851) used starch as his inert substance.

(c) Impregnation in blotting paper

Grazzi (1899) allowed blotting paper which had been impregnated with various test substances to dry and then asked the subject to inspire air which had been drawn through it along a cardboard tube. This tube was closed by a series of cardboard diaphragms with circular openings in the centre. The diaphragms could be progressively removed leaving others with openings of increasing size until the odour threshold was perceived. In this case the minimum perceptible odour is measured in millimeters according to the diameter of the opening which varies between 5 and 50 mm.

(d) Zwaardemaker's olfactometer

Like all Zwaardemaker's work in olfaction this olfactometer is of considerable historical importance as it represents a completely new approach.

The equipment comprises two tubes, one inside the other. The outer tube was made of porous gutta-percha and could be im-

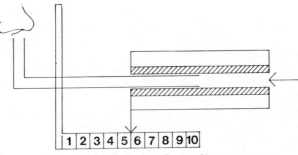

Figure 5.1   Zwardemaker's olfactometer.

pregnated with an odorous substance. The inner tube was of glass and held so that it opened at the patient's nostril. By slipping the outer tube along the inner one a greater or lesser odorant area is exposed to the air which passes into his nostril when the patient sniffs (Fig. 5.1).

The odour threshold then is the smallest surface of the porous tube which emits enough perceptible odour. It is measured in centimeters which represent the movement of the outer tube.

Zwaardemaker suggested that this perceptible minimum represented a useful value and termed it an 'olfactie'. The olfactie could be determined by the mean of the minima perceived by normal people. In clinical practice he used four porous tubes which could produce smell intensities of between 10 to 5000 olfacties.

In order to produce a controllable stream of air, Zwaardemaker later added a pump to his olfactometer and this showed that the odorous threshold was directly dependent on the pressure created by the pump.

(e) Dilution in liquids

Not unexpectedly, many techniques have been offered using odorants diluted by various liquids. These, particularly the use of alcohol, have been criticized as they are often both odorant and odorivector. Some have been of considerable value.

Tolouse (1899) used 54 tubes each containing 15 ml of odorant solutions. He divided these tubes into two series, one for quantitative and one for qualitative use. The quantitative series consisted of 34 tubes containing camphor solutions at dilutions of between 0 and 100 per cent. The qualitative series was only 10 tubes containing: olive oil; orange blossom water; laurel water; violet; rose essence; aniseed; mint; garlic; camphor; vinegar. He used a third series of five tubes containing various dilutions of ether and five tubes of ammonia solutions. This last series was in order to test the common sensation of the nose.

The importance of Toulouse's method is that it introduces very firmly the qualitative as well as the quantitative aspect of olfactory investigation in clinical practice.

A number of clinicians subsequently used a similar approach with these or other odorant substances depending on what each found the most easily obtainable. The most important contribution along these lines has been that offered by Proetz (1924) in improving the technique of progressive dilution.

PROETZ'S OLFACTOMETER

Proetz prepared a rack containing one hundred bottles arranged in ten rows forming a square. Each row represented an odour and each bottle in a row an intensity. He pointed out the importance of the diluent saying that it should be odourless and non-volatile, must not be oxidizable by contact with the air or have any chemical effect upon the substances dissolved in it. For this purpose he employed liquid petrolatum of specific gravity 0·880.

In the selection of odours he chose ten substances, each representing a general class of compounds and arranged in order of complexity, putting the simplest in the first row. He selected characteristic odours which the patient would recognize again, and substances which would not irritate the nasal mucosa and thus declare themselves without stimulating the olfactory sense. Naturally they had to be soluble in oil. The following substances were eventually chosen and Proetz calculated the minimum perceptible amount:

TABLE 2. *Minimum perceptible amounts* (*g/litre petrolatum*)

Idoform	0·00250
Methylsalicylate	0·00297
Amyl alcohol	0·00275
Xylol	0·01730
Nitrobenzol	0·03050
Phenol	0·01921
Guaiacol	0·00302
Cinnamon oil	0·00106
Eugenol	0·00053
Coumarin	0·00077

In order to provide a quantitative element, Proetz decided that a unit of magnitude was necessary. This unit he called an 'olfact' and it represented the minimum perceptible for a substance by many people expressed in grams per litre. Thus 2 olfacts or 3 olfacts was two or three times that amount.

The testing apparatus contained in each row solutions of: $\frac{1}{4}$, $\frac{1}{2}$, 1, 2, 3, 5, 10, 25, 50 and 100 olfacts.

Von Skramlik (1936) built an olfactometer consisting of 49 bottles of odorant substance. These were secured on to a rotating platform and were interconnected by an elaborate system of tubes and taps so that each smell was presented separately to the nose by turning the platform, or could be mixed in varying proportions.

(f) Simple controlled-stimulus methods

Woodrow and Karpman (1917) had attempted to control the physical nature and intensity of the stimulus by using Avogadro's principle. Equal volumes of gases or mixtures of gases contain the same number of molecules if they are at the same temperature and pressure. In that way the number of molecules of an odorant vapour can be calculated. Woodrow and Karpman's technique, using saturated vapours at controlled temperature, pressure and rate, was very valuable as a scientific measurement, but as it could only be used at saturation level, the minimum perceptible thresholds could not be measured. Nevertheless, this can be considered as the precursor of methods which offer a controlled stimulus which can be measured objectively.

Elsberg and Levy were among many workers interested in this approach and in 1936 described their method of *blast-injection*. The importance of this method of presenting the smell to the nose in a controlled manner cannot be stressed too much. From the clinical point of view it has been the major step in the investigation of smell abnormalities.

They used a 500 ml container containing 30 ml of an odorant solution so that the air above it was saturated as in the manner

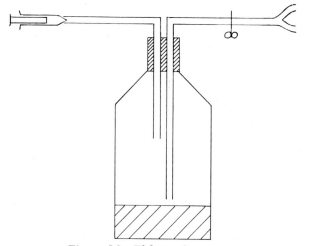

Figure 5.2   Elsberg olfactometer.

of Woodrow and Karpman. Two glass tubes pass through the stopper, one being attached to a syringe and the other to a rubber tube. This tube is presented to the patient's nose and can be occluded by a clip (Fig. 5.2).

The rubber tube is held at the patient's nostril with the clip closed and a volume of air is injected into the bottle, thus raising its pressure so that when the clip is released an equal volume of odorized air is blasted into the nose. The test starts with small quantities of air and the volume of the blast is increased until the threshold is found. Elsberg and Levy's original threshold was that of *identification* of the substance and not simply of *perception*. They also experimented with *stream injection* by attaching the input tube to a compressed air cylinder. They found that, of the factors in the identification threshold, the pressure of the stimulus was more important than the volume of odorant used.

Many criticisms have been made of the blast injection technique. Le Magnen (1945) suggested that the stimulus was artificial and designed a system to overcome this. Wenzel (1948) pointed out that the number of molecules presented by the blast could not be calculated, that the stimulus pressure was inconstant, and that there was a considerable chance of odour contamination of the system. Despite these criticisms, there is a definite value in using this volumetric scale and more clinical use has been made of Elsberg and Levy's method than of any other olfactometric technique.

Le Magnen's modification was based on his objection that the stimulus was presented in an artificial manner. A glass bell-jar of 2 litre capacity was suspended over two concentric cylinders. The space between them was filled with water. The bell was filled with air at a temperature of 20°C and kept at atmospheric pressure by a counter-weight attached to the bell. A vaporizer mixes an odorous substance to the air in the bell and this is precisely done so that the concentration can be determined at that temperature and pressure. The patient breathes in through a tube and the inspiratory force can itself be regulated by a whistle which emits an appropriate sound when the right force has been achieved. This maintains a certain degree of standard conditions between one test and another.

Fortunato and Niccolini (1949) placed six Elsberg containers on a rotating platform. They contained: phenol and musk which provided an olfactory stimulus; citral and guaiacol which also stimulated the trigeminal nerve; vanillin and pyridine which added a gustatory element. The patient breathed in the substances in a normal manner.

The author has modified these techniques further (Douek, 1967), firstly in order to add a qualitative element to quantitative measurement and secondarily to bring clinical tests into a closer relationship with the more recent theories of smell (see Chapter 4).

THE OLFACTORY SPECTROGRAM (o.s.g.)

The author's work is entirely clinical and the first necessity therefore was to find a method which was quick and easily carried out in the out-patients' department or the ward (Figs. 5.3 and 5.4).

Figure 5.3   Olfactory spectogram testing equipment.

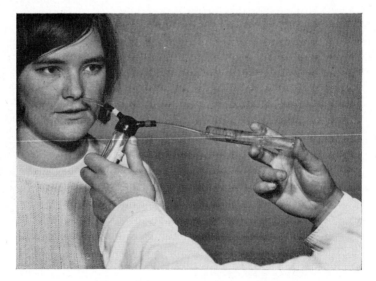

Figure 5.4   o.s.g.—testing a patient.

In diagnostic work and in clinical research of this type, the starting point is to see whether there is any correlation at all between the disease if it is known or at least the complaint and the qualitative or quantitative findings. If no such correlation can be demonstrated then this approach is best dropped. For this reason there is no need here to know absolute values such as the number of molecules in the stimulus and, if the apparatus such requirements demand is complex, then the confusion introduced is counterproductive.

It is well known to the clinician that subjective tests, where the patient has literally to give an opinion on a perceptual phenomenon, are dependent on so many factors of circumstance and mood that it can only be misleading to ascribe any degree of precision to them. If, on the other hand, the crudity of such a test is understood and accepted, then it is of course of enormous value to the doctor to know broadly what the patient feels.

In the final analysis a clinical test has to be judged not on the theoretical considerations which help a scientist understand fundamental issues but on whether it helps a doctor in diagnosis, prognosis and, if possible, management.

Elsberg and Levy's method was quick and simple and had proved most valuable in assessing the neurological cases to which it had been applied. The principles behind it, therefore, were accepted by the author, but the following modifications were made:

1. The minimum identifiable odour did not prove entirely acceptable because the ability to recognize a smell depended on too many cultural and other factors to be standard. The threshold used then is the minimum perceptible odour or the point at which the patient feels he can smell something, even if he cannot identify it.

2. The odorized air is blasted into the nostril and the patient asked to sniff immediately afterwards. This more 'natural' method produced lower thresholds than Elsberg and Levy obtained. 0·5 ml is common among normals and represents the volume deposited into the nostril.

3. It was important to provide a qualitative scale as well as a quantitative one, because abnormalities may lie in that field rather than in quantitative changes. The question arises as to what substances to use.

4. Recently a number of theories of smell, each backed by some scientific evidence, have been proposed (see Chapter 4). It is right

that the qualitative scale should be based on one of these. This is important because, unless we make use of scientific advances, clinical practice does not progress to its fullest potential, and secondly, none of the scientific work has had the benefit of being considered in the pathological situation.

For the reasons explained at the end of the last chapter, Amoore's theory was taken and his seven 'primary' odours accepted as test substances; but it should be clear that clinical tests cannot be asked to prove or disprove the existence of primary odours. The importance of choosing them lies in the fact that they are by far the most common and that this must represent some common feature.

For this reason the following smells were chosen:

1. Ethereal   (E)
2. Camphoraceous   (C)
3. Musky   (Mk)
4. Floral   (F)
5. Minty   (Mt)
6. Pungent   (Pn)
7. Putrid   (Pd)

The actual substances selected were: ether, camphor, phenyl acetic acid, salicylaldehyde, peppermint, formalin and thiophenol. Later these were modified and a synthetic musk used instead of phenyl acetic acid; linalol, which has a very pleasant smell, replaced salicylaldehyde, and laevophenol has a minty smell without having peppermint's cooling effect. Thiophenol, which has a very suitable putrid smell, has caused trouble to some people who used this technique because it is so powerful that only very minute quantities can be used. Other putrid-smelling substances such as skatole or some mercaptans may be easier to handle.

The results can be recorded in block graphs as shown in (Fig. 5.5). Each horizontal column rpresents one of the test substances. The threshold of smell for each substance is represented by the length of the columns. These levels correspond to the amount of air in millilitres injected which is necessary tó produce smell and are marked in 'units'. It would be possible to calculate the number of molecules in every millilitre of saturated air at any particular temperature and pressure for each individual substance, but these values are irrelevant to the purpose of this test and would only introduce confusion. What remains, then, is that each horizontal column

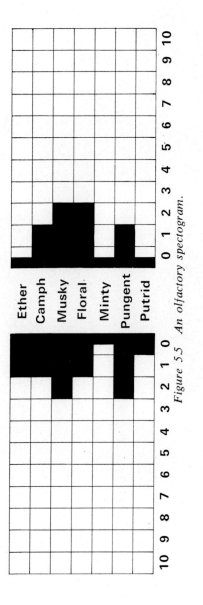

Figure 5.5 An olfactory spectogram.

D

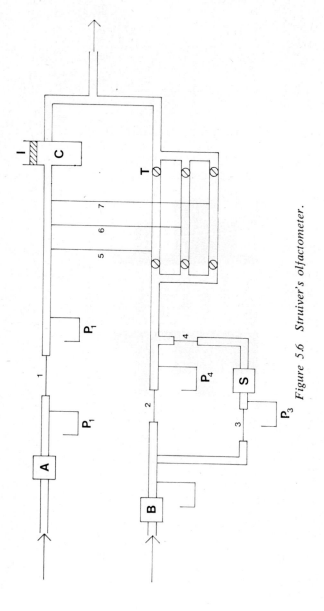

Figure 5.6 Struiver's olfactometer.

is a representation of the Minimum Perceptible Odour for that particular substance in units applicable to that substance only. The other columns become comparable only when the broad term 'units' is used and it would be misleading to give them a more specific name.

This test has proved reliable in clinical practice and has produced remarkably accurate prognoses over the past five years.

2. Complex methods of measuring the minimum perceptible odour

In 1950 Stuiver built an olfactometer which was a good deal more elaborate. By means of a system of two circuits he could control the output of odoriferous air, the duration of the stimulus and also the number of molecules involved. The two circuits are shown in a simplified manner in Fig. 5.6. The first is an air circuit. Purified compressed air enters at A at a maximum rate of 0·9 litre/second. The air is then washed, humidified to 32 per cent and warmed to 17°C. It then crosses a capillary tube (1), the flow across this being calculated from the difference measured at the two pressure gauges P_1 and P_2. The air then flows through a cylinder C towards the nose. This cylinder contains a piston which can be displaced downwards to allow the air to flow outside the system.

The second circuit receives purified nitrogen from a cylinder at point B. This stream flows at 10 ml per second and is divided into two streams. The main flow of 9 ml per second passes through the capillary tube (2) and a slow one of 1 ml per second through two capillary tubes (3 and 4). This stream is saturated with an odorous substance in the container S which is kept at a constant temperature and then passes through the capillary tube (4) to join the air stream. The quantity of odorous substance which reaches the air stream can be altered by using capillaries of different diameters (5, 6, 7) and closing the tap T.

The final concentration of odorous substance is calculated from the air flow; the ratio between the diameters of capillary tubes 2 and 3; the flow in capillary tube 5 which can be obtained from the difference between the pressure gauges of P_2 and P_4; and the vapour pressure of the substance at the known temperature.

The same principle was applied by Bozza et al. (1960), but they used mixtures of odourless and odoriferous air (Fig. 5.7). Air enters at point A, passes through a compressor and into a reservoir, where it is kept at constant pressure by a valve V. From this reservoir a tube leads the air through two circuits.

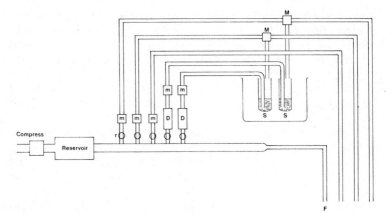

Figure 5.7   Bozza's olfactometer.

In the first instance it may pass through three side tubes, each with flow regulator (r) and flow meter (m), and lead to a large rectangular tube into which the nose and mouth of the subject may be placed. The tube (F) is a continuation of the original reservoir tube and controls the air flow by a valve.

Two other side tubes carry air through drying towers containing silica gel (D) and flow meters (m) into containers where odorous solutions (S) saturate the air in a thermostat. The saturated air then joins the other tubes by passing through mixers (M).

In this way the concentration of the odorous substance can be kept constant while the subject breathes in a natural manner.

Eyferth (1969) has modified Elsberg's method in an ingenious manner (Fig. 5.8). The air emerging from the odorant bottle is connected to a series of capillary tubes of different diameter by means of taps T. The subject breathes in at O a mixture of room air, through the capillaries R, and odoriferous air from the system described. By turning the taps T in such a way as to open progressively the larger tubes the threshold is calculated.

Even more complex systems have been produced by introducing instruments such as the gas liquid chromatograph to analyse the odorant, as few of them can be free of impurities.

3. The olfactory room

A number of attempts have been made to have the subject sitting in a room, or 'olfactorium', into which odours are introduced. The difficulties involved in getting rid of odours once they have been tested are very great and such methods have not been found to have much practical value.

Recently a similar idea was taken up by Kettel (1968). He sat his subject in a box with an opening 1·5 cm in diameter and attached to this a Stuiver type of apparatus; but the air inside the box is completely renewed every minute and a half.

4. Recognition tests

All the tests which have been described can measure the Minimum Perceptible Odour (M.P.O.) and can also be used to measure the Minimum Identifiable Odour (M.I.O.). Indeed, Elsberg and Levy based their original work entirely on the M.I.O. which is a somewhat higher threshold than the M.P.O. Since then most workers have found identification less reliable when measuring threshold but it clearly has a place when testing the higher functions of the brain. Even so, the examiner must be wary because cultural, educational and professional factors have such an important significance.

Sumner (1962) reported an investigation planned to determine how readily conventional test substances could be identified and to

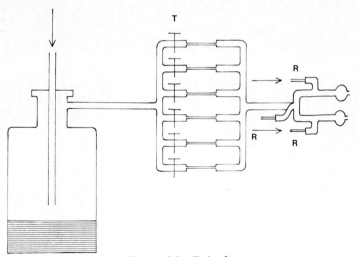

Figure 5.8   Eyferth.

compare these with other substances. He used two hundred subjects who were supposed to have a normal sense of smell. One hundred and fifty of these were out-patients whose conditions were not associated with the sense of smell, and the remaining fifty were doctors, medical students or nurses. The substances which were offered for identification were presented in the following order:

1. Benzaldehyde   5. Oil of cloves   9. Blank
   (almond)   6. Oil of aniseed   10. Nutmeg
2. Oil of lemon   7. Chocolate   11. Tar
3. Camphor   8. Oil of peppermint   12. Oil of eucalyptus
4. Coffee   13. Asafoetida

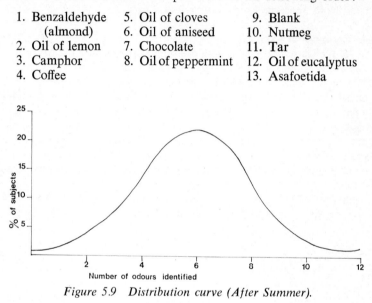

Figure 5.9   Distribution curve (After Summer).

Only two subjects were able to identify all the substances, but the general standard of identification was low. A distribution curve showed that the majority of subjects were able to identify only half the substances used (Fig. 5.9). Asafoetida which was often recommended as a test substance was recognized only by a pharmacologist and a pharmacist.

This demonstrated the problems encountered in identification tests and Sumner's suggestion was that only four substances: coffee, benzaldehyde, tar and oil of lemon should be used, as four out of five people should be able to identify any two and nine out of ten at least one. Chocolate was easily identifiable but presented practical problems in its presentation.

II. OBJECTIVE TESTS OF SMELL

Important as subjective tests are, they cannot provide the type of information that objective tests can offer. It is not as though objective tests are 'better' but rather that they give information of a different nature. Subjective tests demonstrate the rapidity of the mental process, the precision and the ability to understand and describe, as well as perceive, sensory phenomena. This knowledge is important in itself and it is wrong to reject it as somehow confusing and hiding the basic process. On the other hand, it would be useful to separate the two so that each aspect of olfactory perception can be assessed in its proper context.

It is for this reason that a number of attempts at objective testing have been made. Roseburg (1968) classified the following methods:

1. The olfacto-pupillary reflex

It had long been known that many stimuli may produce rapid constriction of the pupil followed by a dilation which can last up to a second. This response has been used in child audiometry but Luchsinger and Brunetti (1948) note a similar reflex responding to olfactory stimuli.

The author has a long experience of assessing both auditory and olfactory abnormalities and has learnt that this test is not only of little value but is positively misleading. As virtually all external stimuli and probably many internal ones produce pupillary changes, conditions have to be remarkably quiet and motionless before a change in the diameter of the pupil can be ascribed to an olfactory stimulu. A similar response can be obtained from a stream of odourless air or as the result of the visual stimulus of the approaching hand of the examiner.

2. Cardiovascular changes

(a) Rise in blood pressure

Unpleasant or violent smells have been known for a long time to cause a rise in blood pressure. This was noted by Vaschide in 1901. This fact however has proved of little value because it is only odours that have a pungent, often painful, element that give consistent results. It is likely, therefore, that it is the common sensation that is involved rather than the olfactory sense and that the reflex is mediated by the trigeminal nerve.

(b) Changes in the pulse rate

Any sensation suddenly introduced may cause an alteration in the pulse rate. In the author's experience this has been of little value as an indication of threshold levels, while at suprathreshold levels of stimulation pulse rate changes certainly occur but are difficult to separate from other causes.

(c) Plethysmography

This has been used by a number of workers. The hand is placed in a plethysmograph which shows an initial short-lived increase in volume followed by a decrease. Nyssen (1933) could only find clear-cut changes in 53 per cent of his cases.

3. Respiratory changes

There is no doubt of the closeness of olfaction to respiration and it is not surprising that many attempts have been made to correlate olfaction to changes in respiratory rate and volume. Most of these are of little value and require mention only for their historical interest.

According to Allen (1929), violent odours diminish the respiratory rate and volume while faint ones enhance it. The reason for this must be protective in the first instance, while the powerful respiratory movements, often excessive, which are shown by a subject sniffing threshold odours are not infrequently observed in the second.

Bourgeois (1933) tried to use these facts in the investigation of legal cases, as very unpleasant smells may even interrupt respiration.

In the author's view, this may be a more fruitful field in which to search for objective responses than others, particularly now that monitoring methods are so sophisticated.

4. Psycholgalvanic tests

The psychogalvanic reflex test is based on the fact that the electrical resistance of the skin is altered as a result of psychic changes. These are quite non-specific, as memories, fear, anxiety, etc. can bring this about. Arrest of the attention by unexpected sensory stimuli sometimes produces marked changes in skin resistance and this technique has been widely used in medicolegal investigation both in lie detection and in patients who claim sensory loss from injury.

According to Bytel and Van Iterson (1925) the skin resistance to a galvanic current falls from 3·400 Ohms to 3·000 Ohms about five seconds after olfactory stimulation. Measurements made with a simple string galvanometer of Einthoven type suggested that there was a correlation between the strength of the stimulus and the amplitude of the deflections obtained.

In 1952 Fortunato described a detailed technique using conditioning to the current which has been of considerable medicolegal value.

The patient is placed in a quiet room free from smells, and lies on a couch with his eyes closed. This minimizes the effect of external stimuli. Two sets of electrodes are then attached to the patient:

(i) Electrodes recording the skin resistance connected to a galvanometer. These are attached to the soles of the feet;

(ii) Conditioning electrodes attached to the calf and which provide faradic stimulation.

An olfactory stimulus is given while the patient breathes normally and two seconds later he receives a faradic stimulus. This cycle is repeated after 10 seconds. the galvanometer showing increasing response to the olfactory stimulus. This type of reinforcement of the olfactory response by conditioning is very effective. If an ammeter is introduced into the circuit the intensity of the current and its variations can be recorded.

Fortunato showed that after conditioning has been obtained the olfactory stimulus can be diminished in strength until it reaches the subjective threshold and that the cutaneous response will be retained almost at that level.

5. Electro-encephalography

Alterations in the alpha rhythm of the e.e.g. can be obtained by olfactory as well as the more frequently used visual stimuli. They are produced in most subjects and for all odorant substances but they have been of little use in olfactometry because only powerful stimuli engender changes which are of practical value (Semeria, 1958).

With the development of Auditory Evoked Responses using an averaging computer and their success in measuring the response to auditory stimuli virtually at threshold level, interest has turned towards the possibility of Olfactory Evoked Responses. A few reports have appeared suggesting that these have been recorded, but none has been consistent or well substantiated. The author's efforts in this direction have demonstrated the difficulty of synchronizing the olfactory stimuli but even more important the difficulty of eliminating the effect of trigeminal stimulation.

Everything points, however, in the direction of early progress in this field.

6. Polygraphic methods

Many workers have associated different measuring techniques such as those described in systems of simultaneous recording. These are certainly valuable, as comparisons can be made between different methods and there is the hope that if a response is missed by one recording system it will be picked up by another.

Of all these techniques, averaged evoked responses are probably the most helpful in the long run.

III. HAEMATOGENOUS STIMULATION

Some attempts have been made at injecting a non-toxic but odorant substance intravenously and then recording its arrival in the olfactory cleft either by the subjective sensation or by objective methods. The measurements that can be made are those of olfactory threshold by diminishing the quantity injected and the time taken before perception occurs. This last is indicative of the state of the circulation rather than of the olfactory sense, and, as a whole, haematogenous techniques have as yet not found an important place either in clinical practice or in research work.

There remains a discussion of psychophysical methods and tests of adaptation which are, to a certain extent, interrelated. They involve subjective sensation and, with the expected progress in that field, objective measurement in the future.

Psychophysical methods

The place of psychophysics in the study of sensory phenomena is now so important that it should be considered separately from both subjective and objective tests.

Psychophysics is concerned with studying the relationship between stimuli and sensations in a quantitative manner. Attempts at this approach started well over a century ago and were quickly dominated by Fechner. His insistence that the intensity of sensation increases as the logarithm of the intensity of the stimulus has obscured the true nature of these relationships under the dogmatic name of 'Fechner's Law'. In many instances it has been shown that this relation is not a logarithmic one at all, rather that the subjective magnitude increases as a *power function* of stimulus intensity.

Fechner also introduced an error of judgement as a unit of measurement (Stevens, 1961) in the shape of the 'just noticeable difference' or jnd. Fechner assumed that error was constant all along the psychological scale which has proved to be clearly mistaken.

Since that time, however, the technology of assessment of the input and output of the sensory system has gradually diminished the effects of Fechner's Law.

Stevens (1946, 1951) has listed four scales for measuring sensory function:

1. Nominal—a determination of equality;
2. Ordinal—a determination of greater or less;
3. Interval—a determination of the equality of intervals or differences;
4. Ratio—a determination of the equality of ratios.

The greatest advances have been in multidimensional investigations. Many different types of odour are tested and described according to as many criteria as possible by many people. The large numbers help to increase the reliability of the descriptions. These subjective impressions can be compared with electrophysiological measurements in animals, such as in Døving's work (1966, 1967). He classified the effects of smells as excitatory, inhibiting or negative according to the electrical changes in single nuerones of the olfactory bulb of a frog preparation. The results were compared with those of judgements in man and these showed relatively high rank-order correlations. In the same way the smells used can be based on theoretical considerations such as the puncture theory or stereochemical theory. The author has attempted to include the effect of pathological changes to the theoretical choice of odour as an extra dimension.

The difficulty lies in interpreting all the factors in such extensive multidimensional operations where complex statistical analyses are required.

Adaptation and its measurement (Supra-liminary olfactometry)

When the olfactory mucosa is stimulated with a smell for a prolonged period, the olfactory sensation gradually diminishes and eventually disappears altogether. This is a well-known phenomenon and it is so marked in smell as well as in taste and touch that it is often given as an example of sensory adaptation. The cause of adaptation has been discussed in Chapter 3, but is still not completely resolved because some animal experiments show that the olfactory system continues to show recordable electrical responses despite prolonged stimulation. This has led Beidler (1957) to suggest that adaptation may in large part be centrally mediated phenomenon. If that is the case, then in the author's view its study in pathological cases should be encouraged because of its possible localizing value in intra-cerebral lesions. The work already done in this field is somewhat limited and will be discussed in the following chapters.

The way in which this is done is as follows. A stream of air saturated with an odorant is directed into the nose and the subject asked to say when he ceases to smell the substance. This is the adaptation time. The stream is then stopped and the subject is then tested with the same substance every half minute until he can smell it again. This is the recovery time.

Adaptation has caused considerable interest among those concerned with psychophysics. Measurements have varied from this clinical test, as tests have been carried out counting the number of inhalations required to produce adaptation, and adaptation times worked out for different intensities. Eckman *et al.* (1967) described the adaptation function by the exponential function:

$$R = a+b/c^T$$

where R is perceived intensity, T the duration of the stimulus, a the asymptote of the adaptation function and $a+b$ the initial value of the function at $T = O$.

Again Eckman and his colleagues (1968) have shown that the perceived intensity of stimuli above the threshold is a simple logarithmic function of duration:

$$R = a+b \log T$$

An important difficulty in tests of adaptation is the expectancy of the patient. This is a problem in all subjective tests, but particularly so when the patient has actually to be told to expect an alteration in perception.

CHOICE OF TEST

To end this chapter an answer or at least an explanation should be found to Wenzel's critical questions.

This review has demonstrated the many weaknesses which are inherent in all the tests of smell. If it is well regulated and controlled the administration of the stimulus may become 'unnatural'. If sniffing or 'natural' breathing is involved, the inconsistency of the duration or force of inspiration introduces other variables. The uncertainty of the degree of saturation of the odorant airstream and the fluctuations in temperature make molecular counts unreliable. When the faint odours of threshold measurements have to be assessed, even the slight differences due to adsorption on to glass or diffusion into the surrounding air may matter.

One important source of error which has received surprisingly little attention was pointed out to the author by Poynder (1968). The odorant substances which are used in so many experiments are more likely than not to contain impurities. Even if these are minute in quantity if analysed, say, in liquid form, it is quite possible that because of differences in vapour pressure the proportion of impurity in the supranatant vapour may be much greater. In other words, even the nature of the odorant itself is in doubt unless the odorant airstream is itself subjected to analysis by, for instance, gas-liquid chromatography.

With all this in mind, the question that must be asked in each particular instance is: What are we trying to measure? Wenzel was critical of the simplicity of many techniques proposed even now that so much technology is available; but the author was recently told of a visit to a centre well endowed with equipment. The patient was subjected to olfactory testing combined with rhinomanometry by a system so elaborate that it occupied a whole room, and the test itself took some hours to perform. The visitor was amused to find that at the end of this complex procedure it was decided to carry out a submucous resection of the septum—a minor operation which can be judged simply by the inspection of the nose with the naked eye.

Clearly workers here are at cross purposes and the choice of test depends on the purpose for which it is required. A very elaborate test for a simple clinical situation is not practical and can be misleading, but the use of a sniffing bottle in an animal experiment of some sophistication is ridiculous. The important thing is to know exactly what is being measured and the limitations of the test.

The author's practice in clinical work is to use a simple collection of sniffing bottles if anosmia is the complaint. If some smell is present, then the Olfactory Spectrogram technique is used so as to assess quantitatively and qualitatively the degree of smell remaining. The results of this simple method have been valuable in offering a prognosis and suggesting treatment. More complex methods must be used in more specific research work.

Although psychophysical methods are not generally suitable in individual cases, the information they produce regarding perception is invaluable.

Finally, the question as to why new methods are produced so frequently can only be answered by the fact that the basis of smell is not yet known. Audiometry remains standard because we know enough about the physics of sound to establish it in firm fact. It could even be said, regarding smell, that new techniques for testing should be introduced each time a new theoretical step is taken. Only in that way can clinical work keep pace with and stimulate research.

CHAPTER SIX

Abnormalities of Smell:
Symptoms and their Investigation

Abnormalities of the sense of smell are many and varied. The descriptions given by the patients are usually heavy with detail of what are considered experiences of an extraordinary nature. This is probably because they involve disorganization of the mechanism of perception and even if the patient's intellect remains entirely normal, it is stimulated by what may be his first glimpse of the world of hallucination and even mysticism. They frequently appear to be more interested in the development of their symptoms than worried by their possible implications.

The clinician may be equally fascinated by the strangeness of the descriptions and it is often difficult to reduce such symptoms to a diagnosis. In practice the minute details of the sensations felt by the patient are often irrelevant and what is important is to extract from the information offered the basic system of abnormalities. In order to do this a simple classification of symptoms becomes necessary and the following arrangement may be found useful:

  (i) Quantitative Changes
              (1) Decreased sensitivity to smells
                  (*a*) Anosmia
                  (*b*) Hyposmia
              (2) Increased sensitivity to smells
                  Hyperosmia
 (ii) Qualitative Changes—Parosmia
              (1) Peripheral type
                  (*a*) Local causes
                  (*b*) Anosmic zones
                  (*c*) S.N.D. response
                  (*d*) 'Essential parosmia
              (2) Central type
                  (*a*) Illusions
                  (*b*) Hallucinations
                  (*c*) Abnormal sense-memory

QUANTITATIVE CHANGES

1. Decrease in the Sense of Smell—Hyposmia and Anosmia

When a decrease in the sense of smell is present it is of paramount importance to differentiate between the cases where there is no smell at all—Panosmia—and those where there is some present —hyposmia. The prognosis will be mainly dependent on such a distinction in many cases. It is necessary to know if the loss of smell is unilateral or bilateral, or predominant on one side, particularly if olfactory tests are to be helpful in localizing intracranial lesions. Partial anosmias, that is the patients who may be unable to smell certain odours and yet have a normal threshold for others, must be distinguished. A test which is both qualitative and quantitative, such as the Olfactory Spectrogram, is essential in order to obtain a clearer view of the extent and meaning of the symptoms.

Although at present it is not possible to suggest tests which can separate mechanical obstruction from end-organ, neuronal or central lesions with the degree of accuracy of auditory tests, it is nevertheless useful to attempt to do so. A combination of history, examination of the nose and the Olfactory Spectrogram can permit a working classification, which apart from its practical use may lead to a less speculative arrangement as further tests become available. The following classification is of value:

Hyposmia and anosmia

Conductive
—Structural abnormalities
—Physico-chemical changes

Perceptive
—End organ lesions
—Olfactory nerve lesions
—Central lesions

Structural abnormalities

These are obvious on examination of the nose. It should be pointed out here, however, that abnormalities should be relatively gross before a hyposmia can be ascribed to them. It is wrong to suggest that a mild deviation of the septum can be responsible for anosmia.

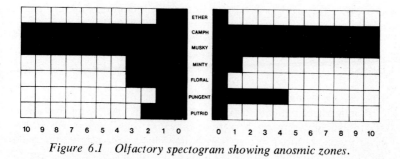

Figure 6.1   Olfactory spectogram showing anosmic zones.

Physico-chemical changes

It was observed (Douek, 1967) that in cases of allergic or vaso-motor rhinitis the hyposmia is frequently so severe that it cannot be explained by any mechanical obstruction. In many of these cases there is in fact very little swelling of the mucosa and certainly not enough to prevent a stream of air reaching the olfactory area. A second finding was the fact that very often the hyposmia was not only partial, but appeared to be selective, affecting certain odours more than others. This was reflected in the olfactory spectrogram by 'anosmic zones' (Fig. 6.1) instead of a general hyposmia (Fig. 6.2). The most severely affected complained of parosmia as well as hyposmia, explaining that smells had 'altered'.

As this feature often varied remarkably in nature and intensity it is most unlikely that it was caused by a lesion of the olfactory nerves or even the receptor cells. The most likely site is in the nose *before* the olfactory cell is stimulated. It is possible that adsorption of odorous molecules may take place in this region, particularly

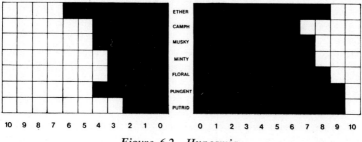

Figure 6.2   Hyposmia.

in view of the changes occurring in the nature and quality of mucus as well as in the texture of the mucous membrane of the nose.

Some experiments to test this hypothesis were carried out and will be described with the 'Anosmic Zones' when parosmia is discussed.

End organ lesions

A number of patients, usually suffering from loss of smell as a result of an upper respiratory tract infection, but occasionally after head injury, report a bizarre symptom. This is often only apparent on close questioning, or even on testing with the o.s.g.

These patients who are unable to smell anything in the normal course of events are aware of an odour when stimulated with over 7 or 8 units of the o.s.g. This smell is quite distinct and always described as sweet and sickly whatever the actual odour. This sensation is short-lived and easily fatigued and can invariably be reproduced with any odour of sufficient strength. Because of the unique nature of the response despite variations in odour, the author has called this the Single Non-Discriminating (S.N.D.) Response until a more exact test can be used.

Any suggestion as to the cause of the S.N.D. response must still be purely speculative but a few factors can be taken into consideration. The fact that nothing of this nature has been described in olfactory groove meningioma despite the careful testing carried out over the past few decades by workers such as Elsberg and Levy, Spillane, and Zilsdorff makes it unlikely that the intra-cranial olfactory system is at fault. The fact that it involves quality of odours implies, in the light of present theories, that the end-organ is affected. In no case has the S.N.D. response been unilateral, and this might militate against an end-organ theory were it not for the fact that no case of unilateral hyposmia of this nature was recorded either. This may be due to the extremely close proximity of the two olfactory mucosae making both vulnerable to the same injury.

If the S.N.D. response can be considered as reflecting an end-organ lesion then it may be comparable to the 'recruitment' present in deafness of end-organ origin.

Olfactory nerve lesions and central lesions

Further differentiation between nerve hyposmia and hyposmia of central origin is extremely difficult if no other neurological signs are present. Usually, however, there are other signs. A useful test here is the measurement of olfactory adaptation or fatigue by the methods described in the previous chapter. The fatigue persists longer in deep-seated lesions which involve the more distal connections of the olfactory fibres. Pressure on the olfactory nerve or on the bulb does not appear to produce this phenomenon but causes simple hyposmia which is often unilateral in its early stages (Fig. 6.3).

Correct identification of odours is also affected more in central lesions, but this involves so many other factors including the previous intellectual status and personality of the patient that it is rarely useful.

There is, however, a large scope for further research in this aspect of olfacto-neurology. The terminal tiers of neurones which carry

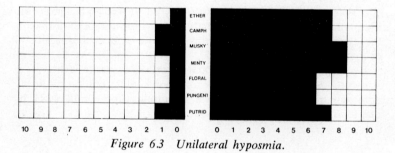

Figure 6.3   Unilateral hyposmia.

olfactory impulses to the higher centres of the brain must serve some function, and they must also be frequently interrupted by lesions in those regions. Although the olfactory changes may be slight and of a nature which is now unknown, their possible localizing features should be considered. The author's present investigations are directed towards identification tests, tests of discrimination between pairs of substances, and their relationship to adaptation.

2. Increased sensitivity to odours—hypersomia

Hyperosmia is a rare complaint and appears to be of psychogenic origin in most cases. This is not an unexpected finding as most sensory organs function at the peak of their capacity and this is accepted as the normal. Any amplification of the experience is difficult to conceive without the intervention of an increased sensitivity of the conscious and interpretative regions of the psyche. A loud

and screeching siren may be a very agreeable sound if it indicates
the end of the day's work while the most beautiful sonata may become intolerable if played repeatedly in the room next door.

The really disturbing end-organ lesions are often associated with
a distortion of the response, such as cutaneous hypersensitivity and
recruitment and pitch-shift in Menière's disease of the ear. In the
sense of smell this would be classified as a parosmia rather than a
hyperosmia.

Some patients describe a particular selective hypersensitivity to
certain odours and not others. This may be a form of parosmia or
it may be due to associative psychological phenomena.

No one has yet convincingly shown a lowered threshold for smell
in these patients or even a physiological hyperosmia during periods
of hunger, during pregnancy or at special times in the menstrual
cycle. The type of temporary hypersensitivity that occurs when
cocaine is applied to the olfactory mucosa immediately before the
hyposmia is of a very special nature and the rapid exhaustion of
the hyper-excitatory state is further evidence against the likelihood
of maintaining such a situation.

Recent work describing lowered thresholds to salt solutions in
low-corticosteroid states will be discussed later, but for ordinary
clinical purposes hyperosmia as a symptom is most unlikely to be
due to an important organic lesion.

QUALITATIVE CHANGES

Qualitative changes in the perception of smells, or the recognition
of smells which no other person can experience are among the most
interesting symptoms produced by disturbance of the sensory
system. Although it is in a field where a clear classification is most
essential, this is seldom done and a state of confusion is general.
The term *Parosmia* is used at various times to mean every possible
aspect of altered smell so that it becomes meaningless. It remains
however so integrated into medical language, that no attempt will
be made here to discard it or re-define it. The term 'parosmia' will
be used therefore to represent all the qualitative changes in smell
whatever their origin or type.

Parosmia can be divided into two groups: a peripheral type and a
central type.

1. Peripheral type

(a) Local causes

These include the patients who complain of a foul smell, when there is ample cause for this. The conditions usually present are infected tonsils, chronic sinus infection, infected teeth and gums and occasionally bronchiectasis. Sometimes neoplasms of this region produce an unpleasant odour in the mouth, of which the patient is conscious. The symptom of foul smell is often referred to as 'cacosmia'.

(b) Anosmic zones—(physico-chemical changes)

A proportion of patients with allergic or vasomotor rhinitis also complain of a distortion of smell. In this case things smell not only different from their expected smell, but also are usually unpleasant. All these cases showed bilateral anosmic zones involving certain specific odours. It should be noted, however, that not all the patients with anosmic zones suffered from parosmia.

It was considered possible that the changes in the nasal mucus and the texture of the mucous membrane were responsible for a selective adsorption of odours. This would mean that in an inhaled odorous mixture certain constituents of the smell would be 'retained' along its passage through the nose, while others would be allowed to go through. The proportion of odorants in the mixture would then be altered when it reached the olfactory mucosa and its smell would be different.

This hypothesis was tested using a series of small tubes, some dry, some lined with a thin film of water, others with nasal mucus. Odorant mixtures were passed through them, and analysed at each end by chromatography. The results showed marked changes in the proportions of the constituents in the mixture (Douek and Poynder, 1970). It is therefore possible that such a selective adsorption of odours may be responsible for the variable and often reversible olfactory symptoms of allergic or vasomotor rhinitis. This is diagramatically illustrated in (Fig. 6.4).

This situation can be compared to a mixture of blue and yellow pigments which look green to the eye. If the stimulus given by the yellow pigment is not forthcoming, then the colour appears blue and not green.

This explanation is given more as a model than a true version of events that cannot be proved, but the evidence in favour of selective adsorption tends to support it.

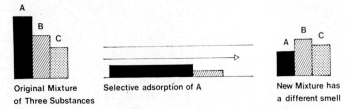

Figure 6.4   Diagram demonstrating selective adsorption.

(c) *The single non-discriminating (S.N.D.) response*

Those cases who show a single non-discriminating response usually present as anosmia or hyposmia, the condition can only be elicited by testing. A few patients, however, point out that they occasionally smell a peculiar sickly smell and have therefore to be classified as parosmia.

(d) *'Essential' parosmia*

In this condition the complaint is that odorous substances with which the patient is familiar have a smell which is quite different from that of past experience. Usually the alteration is for the worse.

If an o.s.g. is taken entirely normal thresholds are found although the patient may not be able to identify the odours. As the cause is unknown, the term 'essential' will be retained.

2. Central type

It is once the peripheral causes of parosmia have been excluded that the greatest confusion reigns. Some clarity has been brought to the situation by Pryse-Phillips (1968) and it is upon his work that the following classification is based:

(a) *Illusions of smell*

These can be divided into two types:

When odours appear to be different from what the patient knows they should be. They are not necessarily bad, just different, and they may vary from smell to smell. In other words the patient may find that a perfume smells of something quite different, which he may or may not recognize. This applies to complex as well as simple odours.

When odours appear not only abnormal, but are all equal and foul. There is no doubt that there is a strong psychological element in these cases, the patients often being greatly disturbed. Pryse-Phillips refers to this type of symptom as 'an illusion of emotional value'. This description expresses the strong emotional content of this symptom.

(b) *Hallucinations of smell*

These can be divided into four types:

Functional hallucinations or synesthesiae. This is a peculiar state in which different types of sensory stimuli can induce a sensation of smell.

True hallucinations. The patient is unaware that he is suffering from a hallucination and believes in the natural existence of the smell he experiences.

Pseudo-hallucinations. In these cases the patient experiences an odour which he knows is a hallucination. This can happen in conditions such as the epileptic aura.

Hypnosis. This is a special situation in which the presence of a smell can be suggested and perceived.

(c) *Abnormal sense-memory*

This is a condition in which a smell can induce the memory of a past smell. The emotional stimulus is so intense that an abnormal volume of forgotten material can be recalled as a result. This can be referred to as the 'Marcel Proust Syndrome'.

Hallucinations of smell are caused by neurological or psychiatric disorders and can be induced by drugs. The o.s.g. is normal if the symptom is of psychogenic origin but it may be abnormal in intracranial disease. In the latter case fatigue and discrimination tests should also be done, as well as other neurological investigations.

HYSTERIA AND MALINGERING

As usual in sensory disturbance these can be very difficult to discover and distinguish from each other. Most cases are connected with a head injury or follow a nasal operation so that claims for damages are often involved. The symptom usually complained of is total anosmia.

(a) The history is often the most important guide, as unless they are well-informed, both these groups of patients claim that they can taste perfectly well, and that it is only loss of smell they are complaining of.

(*b*) the o.s.g. will show no response at all.

(*c*) No response to the chemical-sense stimulators such as formalin and ammonia will be admitted to.

(*d*) In hysteria, there may be an absence of common sensation *inside* the nose but nowhere else in the fifth nerve area.

(*e*) Psychogalvanic, pulse and respiration rate tests are tedious and rarely of clinical value.

INVESTIGATION OF A CASE

It will be seen in the following two chapters that abnormalities of smell occur in such a variety of lesions associated with almost every system of the body. On the other hand these abnormalities occur in a few well-defined conditions with great frequency and only rarely in all the others. This means that some common sense must be exercised when each individual case is examined and the degree of investigation must be appropriate to the history and examination and not follow the total pattern described here.

1. History

The first thing to establish is the nature of the symptom and this is best done by using the system described. If the complaint is quantitative then it is important to know if it is anosmia or hyposmia or more rarely hyperosmia. If it is qualitative an attempt must be made to define its nature carefully and thus to get some indication as to whether it is peripheral or central in type.

An enquiry must then be made regarding taste and questions specifically asked as to whether the patient can taste salt, sweet, sour and bitter, and whether he can taste flavours such as those in food and wines. This is of particular importance in cases where malingering or hysteria are suspected, and where the patient is not sufficiently informed regarding loss of flavour. Occasionally some patients are referred as cases of parosmia and they prove to be real abnormalities of taste and not of smell at all.

The length of time that the patient has been suffering is important and has a marked prognostic value and the nature of the onset is often the most important single feature. Rare cases who seek advice have congenital abnormalities and a family history is necessary. This is usually surprisingly difficult to obtain and in the author's experience information regarding the pre- or peri-natal history is hardly ever available, though questions such as the possibility of maternal rubella may well be of some significance. Other familial illnesses such as epilepsy should be investigated, particularly in the parosmias.

Often the loss of smell takes place during an attack of influenza or at least of severe rhinitis.

If the lesion is the result of an injury there is rarely any doubt, but occasionally it can follow surgery or instrumentation of some type. This last, however, is extremely rare and it is usually an unjustified suggestion.

There should be a careful enquiry regarding associated symptoms. This is particularly important when parosmia is involved because of the possibility of temporal lobe epilepsy. The very obvious features of epilepsy such as automatism, marked disturbance of memory or clear-cut visual hallucinations are not usually a problem, but some less obvious associated symptoms have to be drawn out with some effort. Often the patient suppresses these details out of embarrassment or describes them in a rationalized manner which excludes the extraordinary aspects which would have made the diagnosis clear. The patients who have been referred to the author because olfactory symptoms had dominated the picture produced the following associated symptoms on careful questioning:

(a) Unpleasant taste sensation following a gnawing feeling in the epigastrium rising up towards the mouth and nose, ending as an odour of putrefaction. This was rationalized as 'retching' or regurgitation of evil-tasting and smelling gastric secretion. After close questioning the story was repeated by the patient in such a way that the symptoms were clearly hallucinatory.

(b) Burning odour, with uncertainty as to whether taste was involved, and 'dizziness'. This hallucination of movement was thought to be the result of the unpleasant smell.

(c) Feeling of unreality.

(d) Feeling of anxiety.

If there is any suggestion of mental illness and in particular of depression, these should be noted as well as the personality of the patient.

The patient's general health must be assessed and, more specifically, his ear, nose and throat problems. Any abnormality of respiration, the presence of nose-bleeds, nasal discharge, sinus pain or a past history of sinus disease and nasal polyps are of critical importance. A dental history must not be overlooked but usually patients who have a foul smell and taste in their mouths go first to the dentist and only subsequently consider their symptoms to be due to any but a simple cause. Headaches, if any, must be elaborated on, and questions asked about the patient's vision.

Allergic rhinitis is probably the commonest cause of abnormalities of smell and the history is often quite obvious. Occasionally, however, it is not so clear and a correlation must be looked for between the period of olfactory disturbance and other external factors. Often there are nasal problems like obstruction, sneezing or a watery discharge which point directly to an allergic basis, but sometimes the loss of smell is the most prominent, if not the only, symptom.

Questions must be asked about the patient's smoking habits and any drugs he may be taking or has taken during the period of his olfactory problems. His past illness should be considered and particular note made of suggestion of meningitis, encephalitis or syphilis.

Information must be obtained regarding the patient's work. If it is in a factory, details have to be taken regarding the materials used and about any dusts, chemical or fumes. If the patient works in an office, then questions about the type of heating, the dryness of the air, etc, are important.

In summary, the following points are important in taking the history:

(i) The symptom—nature, whether any loss of smell is total or partial. Whether there is abnormality of taste as well.

(ii) The onset of the olfactory abnormality and the time it has persisted. Association with an illness or injury.

(iii) Associated symptoms—physical, neurological or psychological.

(iv) General health and specific state of the nose and throat and any allergies.

(v) Type of work.

2. Examination

In theory any patient with any symptom could be examined from head to toe and the investigations to which he could be subjected are now virtually infinite. In practice, a reasonable compromise is necessary and a thought out approach is probably more fruitful than general investigations that produce a mass of irrelevant detail.

1. Physical

In a patient with olfactory symptoms the nose must be carefully examined together with the post-nasal space. There should be special reference to structural abnormalities such as a deviated septum, enlarged inferior turbinates and polyps, and to the colour and texture of the lining membrane. The throat and ears are examined as a routine part of an E.N.T. inspection. Signs of infection are important anywhere in that region.

The cranial nerves must be examined and if any abnormality is found there should be a full neurological examination. The ocular fundus should be inspected with an opthalmoscope and the field of vision briefly tested clinically.

2. Olfactory

Initially the patient is asked to sniff bottles containing common substances such as those recommended by Sumner (see Chapter 5). The response of the patient should be classified according to the following characteristics:

No smell at all—recognition not questioned.
Inadequate smell—recognition poor as a result.
Good response to stimulus—poor recognition.
Good response to stimulus—good recognition.

If there is complete anosmia the olfactory tests stop here. If some smell is present, an olfactory spectrogram test is carried out. This will clarify the symptoms considerably, as abnormalities suggestive of local nasal conditions or end-organ lesions will become apparent. Unilateral loss will be recognized and the possibility of an intracranial lesion raised.

If the last possibility is considered, adaptation tests should be carried out. The technique originally used by Elsberg and Stewart (1938), repeated and slightly modified by Semeria (1956) and Zilstorff-Petersen (1953, 1959) is used by the author. Air from a gas cylinder is passed at the rate of 2000 ml/minute through the odorant solution until fatigue has taken place. Then stimulus is presented every 30 seconds until it is perceived. The fatigue commonly lasts between $\frac{1}{2}$ to 1 minute and is not affected by age.

A combination of threshold measurement and adaptation can give some indication of the localization of intracranial lesions, e.g. whether in the anterior cranial fossa or in the temporal lobe.

Other techniques used in the localization of intracranial lesions are now available and of incomparably greater value than the clinical testing of the sense of smell. As preoperative procedures or for confirmation they cannot be replaced, but they are usually elaborate and expensive and often painful. Sometimes there is a morbidity attached. Most patients with smell abnormalities have no intracranial lesions, but if clinical tests suggest the possibility of cerebral tumour they can be useful in deciding which patients should be investigated further.

3. Taste

The sense of taste must always be examined. The first test is with saturated solutions of salt and sugar, sour (e.g. vinegar) and bitter (e.g. quinine or mist. gent. alk.). A drop of the solution is placed on the edge of the tongue, each side separately, and the patient has to recognize the solution.

Taste can be measured quantitatively by means of an electro-gustometer. The electrode is placed on the edge of the tongue and a current measured in micro-amps is induced. The strength of the current is progressively increased until the patient feels a distinct taste similar to that of the pole of a small battery. This is repeated on the other side. The best use of the electrogustometer is in this comparison of one side with the other, as the threshold varies greatly between individuals, between different age groups and between smokers and non-smokers.

3. Ancillary tests

Ancillary tests are of three types. The first, such as sinus and skull X-rays are essential as part of every examination; the second is necessary if certain specific symptoms suggest such a line of approach and this includes skin tests, electro-encephalography, etc. The third group of tests is reserved for cases where serious conditions such as intra-cranial tumours are firmly suspected. The types of investigations available will be discussed first and summarized later, according to their place in the examination.

1. Tests for allergy

(a) Nasal decongestion. A quick and simple test is to spray the patient's nose with a decongestant such as Autistine-privine CIBA (Antazoline sulph. 0·5 per cent, naphazoline nitrate 0·025 per cent) after olfactory testing and then re-test a few minutes after administration. An improvement in the o.s.g. confirms a local, nasal cause with a good prognosis. The patients in which this should be tried are those who have a swollen, congested nasal mucosa and show some degree of smell present with the o.s.g.

(b) Skin tests. The place of skin tests in olfactory abnormalities is even more specific than in other allergic problems. In the first instance it cannot replace a careful history. If such a history is taken in detail, the examiner would have a good idea of the class of allergen in question which can be confirmed by the skin test. A strongly positive skin reaction to grass pollens is irrelevant in a patient who loses his sense of smell only in the water. Allergy to animal hair and dander must not be forgotten while the tests for house-dust allergy are much more valuable since mite-extract has been used.

2. X-rays

This is an essential part of the examination. In the first instance sinus and skull X-rays are necessary.

(a) Sinus X-rays. These may make the diagnosis, as lesions in the sinuses very commonly cause abnormalities of smell.

(b) Skull X-rays. Plain X-rays show abnormalities in only a small proportion of patients. Certain features, however, may help.

Calcification may be present in dermoid cysts or in arterio-venous malformation. Its importance in patients with olfactory symptoms lies in the fact that occasionally meningiomata or a craniopharyngioma may be calcified but often a hyperostosis of the skull occurs in relation to a meningioma.

The vault bones may be eroded by an adjoining tumour or the sella turcica eroded by a pituitary growth.

Occasionally the foramen spinosum is enlarged when it transmits a feeding artery to a meningioma.

The calcified pineal gland may be displaced if there is raised intra-cranial pressure.

(c) Chest X-ray—may show a primary carcinoma of the lung which has metastasized to the brain.

(d) Air encephalogram. This is not an entirely safe procedure if raised intra-cranial pressure is suspected. It is a useful procedure if the cause of a secondary epilepsy is being investigated. It involves a lumbar puncture so that the cerebro-spinal fluid can also be examined for raised protein content and tests for syphilis can be done.

(e) Angiography. Which means injecting a contrast medium into the arterial system to the brain. This is useful in the investigation of space-occupying lesions. It does not matter here if the intracranial pressure is raised, but angiography has certain serious risks of its own.

3. Brain scan

Cerebral tumours may cause a localized break-down of the blood brain barrier and this may be demonstrated by injected substances which have been labelled with radioisotopes. The brain is then scanned and there may be an increased uptake of the radioisotope in the tumour area.

4. Electro-encephalography

An e.e.g. can be recorded even from the fetus and it shows a gradual change as maturation progresses. At first the frequency is slow at 3 to 7 cycles per second (theta rhythm), and then shifts to the alpha rhythm of the occipital regions at 8 to 13 cps and the faster beta rhythm of 14 to 30 cups, although there is a considerable range of normal activity. Sleep produces slow activity which drops to a diffuse 1 to 3 cps or delta activity in deep sleep. Sleep tends to activate any epileptogenic activity in the temporal lobes.

The e.e.g. may be entirely normal in temporal lobe epilepsy and, as it is but rarely that an actual seizure can be recorded, we are dependent for diagnosis on any abnormality that may occur in between seizures. These can be enhanced sometimes by photic stimulation, induced sleep and hyperventilation (Fig. 6.5).

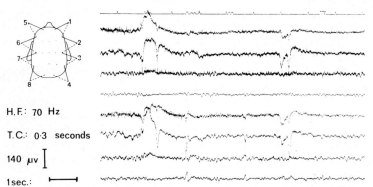

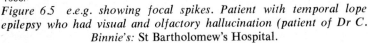

H.F.: 70 Hz

T.C.: 0·3 seconds

140 μv

1 sec.:

Figure 6.5   e.e.g. showing focal spikes. Patient with temporal lope epilepsy who had visual and olfactory hallucination (patient of Dr C. Binnie's: St Bartholomew's Hospital.

Focal epilepsy which arises in the cortex produces discharges over that area, but it spreads to the other hemisphere to produce a mirror discharge on the opposite side. When the epilepsy is of brain-stem origin, it is reflected by diffuse bilateral synchronous discharges over the cortex.

The investigation of a case of olfactory abnormality involves then:

(1) History
(2) Examination—Physical
                —Olfactory
                —Taste
(3) X-ray of skull and sinuses

In cases where allergy is suspected—skin tests are used. In cases where intra-cranial tumours are suspected—electro-encephalography, air-encephalography, angiography and brain-scan are used.

CHAPTER SEVEN

Abnormalities of Smell: Peripheral and General

When the abnormalities which can alter one of the senses are discussed an important difficulty immediately becomes apparent. The symptoms the patient complains of are very similar whatever the cause. The differences that may act as a guide to diagnosis were analysed in the previous chapter but it became quickly clear that seemingly unrelated symptoms and particularly signs elicited by examination or special tests were essential in reaching a conclusion.

The causes of these abnormalities and their development have now to be reviewed, but as intra-cranial lesions belong to a group so separate both in the management of the case and its importance, they are best studied separately in Chapter 8.

No attempt at classification can pretend to approach the true nature of disease and it can only offer a framework for the description of related causes. In this chapter, therefore, the peripheral and general causes of olfactory abnormalities will be studied under four headings:

> Lesions of the nose;
> Lesions of the olfactory nerves;
> Physiological and hormonal changes;
> Congenital and hereditary.

As far as the management is concerned, a broad view of surgery and treatment will be given; but those techniques practised by the author, or taught to him by colleagues and which have not previously been recorded, will be described in more detail.

LESIONS OF THE NOSE

1. Structural abnormalities

(a) Deviated Septum

The most common abnormality found in the nose is a deviated nasal septum. The cause of this is sometimes obvious as in patients who have a fractured nose, but in most cases the reasons are obscure. The main symptom is nasal obstruction, but the interference with sinus drainage which results may certainly be a factor in repeated sinusitis and sinus headache. Anosmia is usually given as a symptom but this is hardly the case. In the first instance the deviation would have to close off both olfactory areas from the air-stream, and this is against the natural effects of buckling of the septum. The author carried out experiments on the air-stream in such a nose using the plastic transparent model and water as described in Chapter 2. These showed that it was extremely difficult to *prevent* part of the stream reaching the olfactory area. To do so required the most unlikely buckling of the maleable septum. These theoretical considerations are reinforced by clinical observation among a large number of patients with deviated septa. Hardly one, seen by the author, suffered from anosmia. On the other hand, a certain number of patients, particularly among those where the deviated septum had contributed to abnormalities of the mucosa, produced minor degrees of hyposmia.

The operation of choice to relieve the symptoms of a deviated nasal septum is its submucosal resection. This operation should rarely be offered, however, as a proposed cure for anosmia. It may be carried out as part of general corrective procedures in the nose and if some hyposmia is present an improvement can be expected. If anosmia is present, then it suggests a cause other than the position of the septum.

(b) Weakness of the alae nasi

This is a less uncommon abnormality than may at first appear. The nose is narrow and the musculo-cartilaginous wings of the nostrils thin and weak, so that with each inspiration they are drawn in towards the septum, thus obstructing the airway. This abnormality is sometimes congenital, but is often associated with a deflected nasal septum; possibly the excessive inspiratory efforts contribute to some extent to the weakening of these structures. The end result, however, is the typical sucked-in appearance which the nostrils take on at each inspiration, thus diminishing yet more the already narrow airway.

Rarely this problem is a major one and elaborate plastic procedures are necessary. Their description is outside the scope of this book; more commonly the condition is mild and an operation taught to the author by H. J. Groves is often helpful.

E

The patient has swabs taken from his nose for bacteriological investigation and if necessary an antibiotic ointment is smeared on the inner surface of the nostrils on the pre-operative days. The operation is best carried out under general anaesthesia. An insulated diathermy needle with its uncovered point triangle-shaped is used as in sub-mucosal diathermy of the inferior turbinates. The sharp point is pushed through the lining of the nostril so that it lies in the sub-cutaneous tissues at about the corner between the nasal bone and the maxilla. A diathermy current of moderate strength is then passed through the needle inflicting a burn in the tissues surrounding the point. The subsequent scarring and fibrosis beneath the skin in this region tethers the nostril in that position.

2. Nasal polypi

These growths are formed from very oedematous and hyperthrophied mucous membrane usually in the region of the ethmoid cells. They are often translucent but become opaque particularly when they are near the exterior. They are clearly not neoplastic in origin and their appearance is usually associated with infection, allergy or both. How these conditions give rise to polyps is not clear but it may be that many forms of irritation can cause the delicate nasal mucosa to swell in this manner.

The two main symptoms in this condition are nasal obstruction and anosmia. Examination of a patient who presents in this manner is imperative and on looking into the nose the unmistakable appearance of the polyps makes the diagnosis obvious.

The treatment of nasal polypi is by excision with a snare and nasal forceps. This can be done under local or general anaesthetic and they should always be sent for histological examination to exclude malignancy.

Regarding the return of the sense of smell, the prognosis in most cases is good, particularly if the patient is treated with antihistamines after the excision.

3. Allergic and vasomotor rhinitis

Nasal allergy falls into groups: seasonal and perennial.

Seasonal allergy is the well known hay fever. It is due to a specific sensitivity to certain pollens and appears from March to September depending on which pollens are the sensitizing ones. During exposure there is itching of the nose, naso-pharynx and eyes progressing to intense irritation, rhinorrhoea, sneezing, and lacrimation. Examination of the nose shows usually a pale blue swollen mucosa with much clear mucoid secretion. The diagnosis here is seldom in doubt and although anosmia is an imporant symptom it is never a problem in itself, as it is as temporary as the condition which caused it.

Although perennial nasal allergy is basically the same disease, there are certain important differences. The allergens are often multiple and may be ingestants as well as inhalants. Loss of smell is a common symptom and patients will often complain that they have not been able to smell anything for years. The clinical picture is not as acute as in hay fever and the symptoms vague. They consist of nasal obstruction, sometimes worse than at other times; a chronic nasal discharge resulting in a drip at the back of the nose. The olfactory symptoms are very important to define because, although the initial complaint is anosmia, it becomes clear when the history is carefully taken that this is neither complete nor unchanging. This feature of fluctuation in degree is most important as it differentiates between those cases which are potentially reversible and others where there is no smell at all and which are not. A few of these cases suffer from parosmia, and the olfactory distortion is always for the worse. Again, it fluctuates in intensity and can sometimes be very persistent.

On examination the nasal mucosa is boggy and bluish in appearance, coated with strands of mucoid secretion. Often there is a super-added infection with mucopus in the nose.

The o.s.g. is a valuable test in these cases. Total anosmia is hardly ever found and the varying degrees of hyposmia are well demonstrated (Chapter 6, Fig. 6.2). Among the patients suffering from parosmia a large proportion show anosmic zones (Chapter 6, Fig. 6.1).

Investigations in these cases start with the history, as some attempt at pin-pointing the allergen must be made. Among the inhalants, house-dust is a very common allergen, animal danders, hairs and feathers follow. Very often more than one substance is responsible. Among the ingestants, milk, eggs and chocolate are sometimes responsible and a number of drugs have also been found to cause such symptoms.

Specific allergy tests such as cutaneous reactions or nasal applications are not as helpful as when dealing with hay fever. Elimination tests have been devised for ingestants. The diet starts with a few foods to which the patient is unlikely to be allergic, and then further foods are added until a reaction occurs. This is seldom practical but an allergen-free chamber into which inhalant allergens are gradually introduced is an even less likely system.

Most commonly the diagnosis has to be made from a combination of history, examination, o.s.g. and skin-tests.

There exists a group of cases where the patient complains of very similar symptoms, namely nasal obstruction, clear discharge and hyposmia. The nasal mucosa appears infected and purplish. There is no history which could incriminate an allergen, indeed the evidence tends to the contrary, and skin-tests produce unhelpful results. The o.s.g. is similar to that of the typical allergic rhinitis. Most patients will complain that the symptoms can occur in a number of varied situations including conditions of stress. The name 'vasomotor rhinitis' has been given to these cases, but from the olfactory point of view they behave in a manner similar to that of allergic rhinitis. In both these types, the loss of smell and indeed many of the other symptoms will improve considerably with anti-histamines (Douek, 1970) thus suggesting a superficial lesion un-likely to affect the end-organ or intra-cranial structures.

4. Infective, atrophic and specific rhinitis

Inflammations of the nasal mucosa, or rhinitis, may be acute or chronic.

The most common type of acute rhinitis must also be the com-monest illness of all—the common cold. It is caused by a number of different viruses and it is very infective, being transmitted by drop-let as well as contact.

During the severe phase of nasal obstruction due to venous stasis, where there is often a secondary bacterial infection, the sense of smell is lost completely. It gradually returns during the following stage of resolution provided there is no complication such as sinusitis or an iatrogenic rhinitis caused by irritant medication. An interesting feature is that many people such as perfumers, who are dependent on olfaction for their work, have noticed that there is a period of hyperosmia during the short prodromal and early irrita-tion stages of the common cold. It is likely that during that period the end-organ or olfactory nervous fibres are themselves involved in the disease process and it is the possible explanation for those cases which develop anosmia following an upper respiratory tract infection.

Rhinitis occurs very frequently with influenza and these are the patients who may lose the sense of smell altogether. It is not possible to know what proportion of patients with influenza lose the sense of smell, as the illness is not reported but the numbers must be very small indeed. The pathology involved has not been firmly established, but one patient at Guy's Hospital has been studied. The infective nature of her rhinitis is not in doubt and she will be discussed later.

Acute specific rhinitis due to diphtheria, syphilis, etc., are now so rare that the author cannot report their effects on the sense of smell.

Chronic rhinitis is usually the result of chronic sinus infection. It usually produces a hypertrophy of the nasal mucosa and hyposmia is common. Partly it is due to the superficial effects of nasal obstruction and overproduction of nasal secretions, but Fig. 7.1 shows the electronmicrograph of a biopsy taken from such a patient. The olfactory cells have been replaced in large parts by a respiratory type of epithelium.

Atrophic rhinitis which is rare in Britain sometimes occurs following chronic infection, but its aetiology remains obscure. The sense of smell is often diminished but rarely lost completely.

Of the chronic specific inflammatory conditions of the nose such as syphilis, sarcoidosis, scleroma and tuberculosis the sense of smell is affected only in so far as the local condition becomes obstructive, productive or secondarily infected. One case of advanced leprosy examined by the author had considerable peripheral nerve involvement and destructive lesions of the nose itself, yet had normal olfaction.

It is worth noting here that the numerous intra-nasal medications taken for the above conditions not infrequently produce a rhinitis medicamentosa. This can be very severe with considerable loss of the sense of smell.

It is traditional to say that abuse of tobacco, snuff and perfumes will damage the sensitivity of the nose and there is no doubt that excessive smoking will have this effect. The degree, however, is difficult to assess and the admonishment is often seen as a recommendation in favour of 'good living'.

Loss of Vth cranial nerve sensation may alter the 'texture' of the stimulus and lead the patient to a complaint of some alteration of perception, although the olfactory sense is itself unchanged.

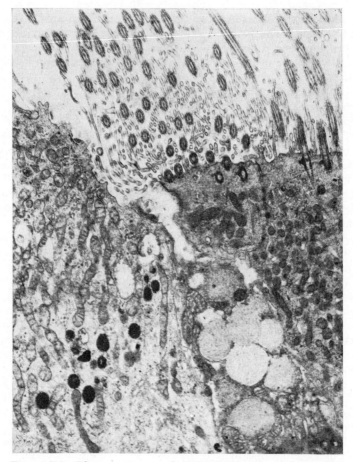

Figure 7.1   Electronmicrograph (×20,000) made at Guy's Hospital *by L. H. Bannister and H. Philipp. Patient of the authors
with chronic rhinitis. Olfactory epithelium replaced by abnormal
respiratory-type cells.*

5. Tumours—olfactory esthesioneuroma

Nasal tumours may be benign or malignant. If they obstruct the air-stream, then the sense of smell may be lost. It is unusual, however, for the patient to present complaining of loss of smell.

There are tumours, however, which originate directly from the neuro-olfactory tissues. They were thought to be exceedingly rare, but in recent times numerous case reports have been appearing with some regularity, possibly because of better histology. Indeed, the diagnosis must be histological because these tumours, known collectively as olfactory esthesioneuroma after Berger's description in 1924, are very difficult to distinguish macroscopically from simple nasal polyps. They arise rather high up in the nose and are perhaps more vivid in colour. There is still not complete agreement regarding the different histological types which comprise these tumours, but Guerrier (1969) divided them into four groups:

1. Esthesioneuroepithelioma,
2. Esthesioneuroblastoma,
3. Esthesioneurocytoma,
4. Esthesioepithelioma.

There is no doubt, however, that there are many transitional forms and sometimes the different cellular arrangements may co-exist. The true origin of these tumours is uncertain and Willemot (1971) pointed out that there were a number of possible sources: the neuro-epithelial cells of the olfactory membrane; the olfactory placode; the vestigial organ of Jacobson; the sphenopalatine ganglion; Loci's ganglion; the central nervous system.

These tumours are all of low-grade malignancy with a marked tendency to local invasion but not metastasis. On the whole, the esthesioneuroepitheliomata are more radiosensitive, but the recurrence rate is higher than esthesioneurocytomas. Lewis *et al.* (1965) described eighteen cases. Only one-fifth of the cases produced metastases. Half survived for more than five years, but recurrence was likely.

The treatment should be wide excision, followed by radiotherapy.

LESIONS OF THE OLFACTORY NERVES

The fragility of the physical barrier between the olfactory mucosa and the brain has been demonstrated in the chapters on anatomy and cranial injury. The fact that the olfactory cilia are bathed in the nasal mucus while the proximal end of the same cell lies in the substance of the olfactory bulb, has for a long time been considered significant in the transmission of disease to the brain.

De Lorenzo (1970) has demonstrated this neuronal pathway by following the migration of ferritin molecules from the nasal mucus of an animal with the electron microscope. These molecules appeared in the olfactory bulb within two hours.

A number of diseases caused by neurotropic viruses are believed to be transmitted in this way, particularly when there are upper respiratory tract symptoms in the prodromal illness. It is therefore likely that this pathway may be taken by viruses, and indeed, other toxic substances which give rise to olfactory disturbances.

Diphtheria toxin may affect olfaction in this way, but it is so rare today that little information is available.

Virus illnesses are sometimes incriminated but usually with little proof. A group of patients exists, however, who give a history of upper respiratory tract symptoms such as sore throat, nasal congestion and rhinorrhoea, together with malaise and sometimes pyrexia. In other words, an illness lasting a few days which would in normal circumstances be put down to influenza. This illness is followed by loss of smell, and the name of *influenzal anosmia* is given to it. It is an important concept as it is the diagnosis most commonly made in cases of olfactory loss not associated with head injury, and some salient points should be considered.

1. There is no way of knowing what proportion of patients suffering from an influenza-type illness lose the sense of smell. These illnesses are, of course, not notified and in most cases treated by the patient symptomatically. As they are probably the most common type of illness encountered, the percentage which show olfactory disturbance must be minute.

2. In most reported cases there can be no certainty that the illness was influenza, or that if it was, it was indeed responsible for the olfactory disturbance. Virological studies are rarely carried out, and if they are, no evidence of a specifically olfactotropic virus has emerged.

3. The reports of cases that are available show considerable uncertainty regarding the subsequent history and the prognosis. The majority of cases described have complete anosmia and few workers doubt that the chances of recovery are slight. On the other hand, there are clinicians who mention that 'some' patients, particularly those whose loss of smell was 'not severe', do have a chance to re-

cover. There is no way of knowing whether these cases are in any way aetiologically connected with the more severe ones that do not recover. It is possible that they are simply patients with rhinitis of one or other cause with an entirely normal olfactory surface.

4. The confusion regarding aetiology has led in the past to even more confusing reports regarding the value of various types of treatment. Use has been made of strychnine on obscure grounds; of steroids, which may be effective in cases of allergic rhinitis and possibly in reducing oedema of viral infections; of vitamin A on ingenious pigmentary grounds. None of these treatments of *olfactory* abnormalities can be properly assessed as it is not clear what the causes were. Thus success with steroids in a patient with allergic rhinitis is no victory when we are concerned with head injuries.

The only pathological evidence regarding olfactory loss following influenza is a case of the author's (Douek, Bannister and Philip, 1973). The patient had an influential illness which she caught from her sister who, in turn, had caught it from a neighbour. All three lost the sense of smell. There is no doubt of the infective nature of the illness and the probability is of a virus infection. Clinically they had influenza. A scrape biopsy of the olfactory mucosa showed severe loss of neuronal elements of every sort in the sub-epithelial tissue as well as loss of surface element (Fig. 7.2).

This makes it likely that influenzal anosmia does exist and that it can cause severe loss of smell. We know that neuronal elements are damaged, but we know little else.

The author's management of such a case, after having excluded other cranial nerve abnormalities, sinusitis, etc., is to classify the olfactory symptoms into: anosmia, hyposmia or parosmia. The prognosis in hyposmia is good, as improvement is likely. Parosmia is more problematical and an olfactory spectrogram is necessary. If the o.s.g. suggests the likelihood of local causes, then the prognosis is more hopeful. When the patient complains of anosmia, an o.s.g. is essential as it may demonstrate that there is more smell present than the patient suspected.

These patients require a prognosis and are usually very distressed by their loss because of their enjoyment of normal olfaction previously. It is wise to offer hope to those who have some residual smell, particularly with an o.s.g. suggesting an acute rhinitis, but to be more guarded with the anosmics. It is always justified to prescribe a course of antihistamines for a period of two months

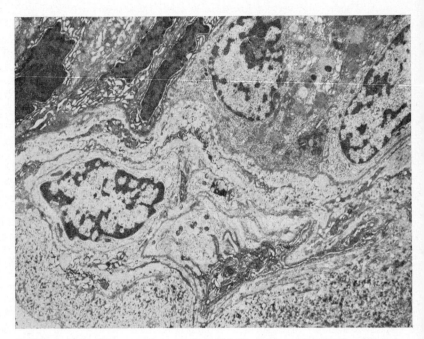

Figure 7.2   Electronmicrograph (×8,000) made at Guy's Hospital *by*
L. H. Bannister and H. Philipp. Patient of the author's, showing loss
of olfactory epithelium with neuronal elements replaced by collagen
tissue.

or more as so many of the local nasal problems respond to these
drugs. It is often surprising how many patients respond to this
treatment and the remaining group who can properly be called
influenzal anosmia is quite small. As has been pointed out in a
previous chapter, when a single Non-Discriminating response is
obtained when testing with the o.s.g. the prognosis is very poor.

Drug-induced anosmia is a subject which has received little attention. When drug-induced hearing-loss is discussed, the point is
always made that the damage is in the organ of Corti and not in
the succeeding nerve fibre. This approach is irrelevant as far as
olfaction is concerned, because the end-organ and nerve fibre share
the same cell, but it may explain why otoxic drugs apparently do not
cause anosmia. It is interesting in this context to reflect on the

extraordinary specificity of ototoxic drugs, as this may suggest a line towards their mode of action.

Certain drugs applied to the nose, such as cocaine and thyrotrycine, have been incriminated, but few such cases exist. In particular, cocaine solution is used very widely to produce local anaesthesia and anosmia never occurs. Cocaine induces a heightened sense of smell followed by a rise in the threshold which returns to normal after half an hour or twenty minutes. It is likely that the belief that cocaine caused anosmia comes from the cocaine addicts who use concentrations and quantities quite different from those administered by rhinologists.

Recently the effect of certain antibiotics and chelating agents on the sense of taste has been discussed (Naess, 1971) and in many instances administration of zinc, nickel or copper has restored the situation. There has been some suggestion that this situation may also involve smell but there is as yet no conclusive evidence.

Loss of smell from occupational contact with vapours is not uncommon and includes various paints, tars, heavy metals, and petrol derivatives. Recently ethylamine, cresols, formalydehyde, sulphur dioxide, carbon monoxide, furfurol and many organic solvents have been incriminated.

It is said that patients suffering from severe diabetes may present an elevated olfactory threshold.

Vitamin A has been given as a treatment of anosmia in the belief that the yellow pigment of the olfactory mucosa contained carotenoids. There is little evidence that vitamin A is of value and this treatment has lost some of its attraction since the discovery that the olfactory pigment was not mainly carotenoid in origin. It is possible that a raised olfactory threshold occurs in vitamin B deficiencies but there is no evidence that administration of these vitamins in other circumstances is of any help.

Olfactory changes occur with age. Although they are physiological there is a tendency today to consider the changes which come with old age as pathological, and in this way many can be reduced or slowed down if properly understood and treated.

It is important to realize that olfactory changes probably occur throughout life, and that they are complex in nature. Darwin observed the lip and eye movements of his baby son and felt that at thirty-two days he recognized his mother's breast. This is mainly due to olfaction. It is difficult to test the sense of smell immediately after birth, but observation of the heart rate and respiration sug-

gest that the newly born will respond to many odours. The author
has presented many mentally retarded children with odorous substances when assessing their response to sensory stimuli, and there
is rarely any doubt that they have smelt them; usually they produce
a mouthing movement and occasionally attempt to put the substance
into their mouths. The association between olfaction and the alimen-
tary tract is very close in children. Kussmaul, over a hundred years
ago, observed the responses of sleeping infants; they purse their
eye-lids, make faces, and move, occasionally waking to a strong
smell. They adapted rapidly on repetition. Fatigue has been studied
in relation to age and according to Zilsdorf-Pedersen it does not
alter.

The threshold itself cannot be measured accurately in children,
but there is certainly a diminution in the sense of smell in the old.
This is a slow involution which according to Chouard (1968) is due
to a reduction of the number of neuro-sensory elements.

With increasing years there is a very interesting phenomenon
of change in olfactory preferences. It is everyone's general experience that children like different foods from adults. Many
attempts have been made to grade the preference for odours by
people of different ages and there is no doubt that marked differences are present. Their exact significance, however, is not clear,
nor the relative roles of experience, memory, nutritive requirements
or hormonal changes.

PHYSIOLOGICAL AND HORMONAL CHANGES

The close relationship between sex and olfaction has been the
subject of great interest and study ever since Le Magnen began to
publish a series of observations on this subject (1948, 1949, 1950,
1951, 1953, 1961, 1963). The importance of this work cannot
be exaggerated and it has been enhanced by the discovery of pheromones in insects and their possible role in animals by Parkes and
Bruce. The physiological aspects of this question deserves independent study, and this is given in Chapter 10.

The pathological aspects are less well known. The mode of action
that an odorous substance may have is made even more obscure
because of its effects on the non-olfactory mucosa of the nose.
Thus, castration experiments have produced a hypotrophic effect on
the whole nasal lining and not exclusively on the olfactory mucosa.
When oestrogens are given, they produce a hyperaemia of the

respiratory mucosa and progesterones a turgescence of the erectile tissues in the turbinates. Clinical observers have many times noted the occurrence of epistaxis at the times when periods are expected. Another feature which may add to the confusion is the possibility that hormonal changes produce changes in nasal secretion and that this, in turn, alters olfaction.

Pregnancy is the state which has produced most comment. This is because of the well-known changes in dietary preferences which occur during its development. Many experimenters have tried to measure the olfactory threshold with the contradictory results which have become only too expected in olfactometry. It is clear that the gross type of measurements available at present are inadequate for a demonstration of the fine qualitative responses which take place. There is, however, one substance which has demonstrated conclusively altered responses and that is known as 'Exaltolide'. It is an odorous substance, a lactone synthesized by Ruzicka and we refer to it again in Chapter 10. It is not smelt equally by men and women, and in the latter the threshold varies with the menstrual cycle. In pregnancy the ability to smell Exaltolide diminishes during the first two months. At the end of the third month it regains its normal level but continues to increase until the end of gestation.

Guerrier et al. (1969) describe six women who had lost their sensitiveness to Exaltolide following hysterectomy and ovariectomy. In all cases these patients regained their ability to smell this substance with hormonal replacement therapy.

The sense of smell of certain cases of hypogonadism secondary to hypothalamo-hypophyseal dysfunction has been studied by Cluzel (1964). He found a decrease in the sense of smell in patients with gigantism, Froelich's syndrome and Simmonds' disease. The most remarkable associations between olfaction and hypogonadism are found among the hereditary and congenital conditions which are discussed in the next section.

There have been a few reports that olfactory acuity increased with hyperthyroidism and decreased with hypofunction of the thyroid gland. Again it is not clear whether this is due to a central hormonal action on the olfactory pathways or to the changes produced in the nose by myxoedema. Chavanne (1937) in particular found that thyroxine inhibited nasal secretion.

Bieber (1959) described anosmia in women hypophysectomized as part of the treatment for breast cancer. Although it is difficult to be sure that this was not a local result of the surgical intervention, he maintained that they recovered following injections of oestrogen.

Thiellement (1955) found that injections of testosterone increased objectively olfactory acuity in the dog. This finding has led to the use of testosterone in an empirical manner as a treatment for anosmia. It has also raised the question of an association between male impotence and anosmia, although no scientific study has been carried out.

There have been some reports that there is an association between adrenocortical insufficiency and an increased ability to detect salt solutions (Henkin and Kopin, 1964). It is the chlorine gas in the vapour which is smelt. The threshold returned to normal after treatment with large doses of steroids though not with DOCA.

Altogether it seems that an association between disease of the endocrine glands and olfactory changes is indisputable. It is also likely that hormonal replacement therapy will restore the situation. On the other hand it is imperative to demonstrate the deficiency which must have more signs than the purely olfactory ones and to treat it with the right hormone. There is no place for empirical treatment with steroids, oestrogen, thyroxine or testosterone in the absence of other indications of dysfunction.

CONGENITAL AND HEREDITARY

As we go to press, new associations between olfaction and heredity appear daily so that no attempt can be made at covering the whole field. It suffices to note the great importance which such a study can have to the development of genetics. We will here divide the better known into groups.

1. With naso-rhinencephalic abnormalities

The author has had the opportunity of studying four patients with posterior-choanal atresia who were under the care of W. G. Edwards. They had no perception of smell even if stimulated from the anterior nares. One patient who had unilateral atresia was operated on and obtained some smell, although nowhere near the threshold of the other nostril. Two cases of posterior choanal atresia who died soon after birth from other abnormalities were found at post-mortem to have arhinencephaly.

Demorsier (1955) saw two different types of dysplasia of the olfactory lobes. In the first it is associated with hypogonadism and will be discussed in the following section, and in the second, telence-phalosynapsis, the cerebral hemisphere are unseparated.

2. Associated with hypogonadism

In Demorsier's olfacto-genital dysplasia there is an agenesis of the olfactory lobes and hypogonadotrophic hypogonadism. Clinically the male presents with anosmia and eunuchoidism, occasionally there are skeletal abnormalities such as scoliosis and genu valgum. There is a reduction in the 17-ketosteroid level and an increase in oestrogens. Although aspermia is present, the chromosomal sex is normal though lacking in sex chromatin. Pathological examinations have demonstrated a hypoplasia of the hypothalamus as a whole together with olfactory loge agensis. The testicles are atrophic.

The female is small, anosmic and has primary amenorrhoea. There is an infantile vulva and atrophic uterus. The ovaries are immature but again the chromosomal sex is normal and sex chromatin reduced.

It is probably transmitted as an autosomal dominant by un-affected or anosmic females.

There is a similarity with a syndrome described by Kohne (1947) where aplasia of the olfactory nerves and bulb is associated with primary eunuchoidism.

This type of hypogonadotrophic hypogonadism may also be associated with cerebellar ataxia.

Turner's syndrome where female hypogonadism with a single X-chromosome has been studied from the olfactory point of view and show normal smell though they do not respond to Exaltolide. The same applies to Klinefelter's syndrome.

In the rare cases of feminizing testicle where there are no chromosomal abnormalities and the genotype is XY but the pheno-type is female, Cluzel (1964) found the response to Exaltolide is also female. This confirms the hormonal rather than genetic dependence of that response.

3. Associated with hypogeusia to hydrochloric acid

Henkin and his associates found patients with familial dysauto-nomia who could not detect salt solutions or hydrochloric acid, but who respond to subcutaneous methacholine injections.

Another syndrome which associates hypogeusia and hyposmia includes submucous clefts of the dorsal hard palate; facial hypo-plasia and retardation of physical growth below the 3rd percentile.

On the other hand, an increased sensitivity to these solutions was present in patients with cystic fibrosis as well as their mothers and male siblings.

4. Associated with pigmentary abnormalities
Three members of the same family, all red-haired and anosmic, were described by Brown (1968), but there may be explanations other than a pigmentary one in those cases.

5. Complete and partial anosmias of genetic origin
Many families where congenital anosmia has appeared in a number of individuals of both sexes have been described, but the most interesting ones are those where anosmia for one or more modalities is present.

A family in which certain individuals could smell only the pink variety of verbenas while others smelt only the red was described by Blakeslee and Salmon (1931). Specific anosmias to many substances such as benzaldehyde, hydrocyanic acid and nitrobenze have been described. Amoore has drawn attention to the value of such discoveries both to genetics and to the progress of olfactory theory.

6. Refsum's syndrome
This is a rare disorder where a chronic polyneuritis with occasional remissions is associated with distal wasting and paresis; ataxia with cerebellar signs; pupillary abnormalities; deafness; ichthyosis; atypical retinitis pigmentosa and night-blindness with concentric constriction of the visual fields and a raised cerebrospinal fluid protein. It occurs between the first and the fourth decade and more than one member of a sibship with normal parents can be affected. There is usually a very high degree of consanguinity which indicates a rare recessive gene. The sense of smell may also be impaired.

CHAPTER EIGHT

Abnormalities of Smell: Intra-cranial Lesions

By and large, if we exclude psychological or psychiatric illness for separate consideration, the intra-cranial lesions which cause abnormalities of smell are of three types:

Trauma
Tumours
Epilepsy

Of these three, epilepsy may well be secondary to a tumour or to trauma, but as the symptomatology is so close to the other epilepsies where no tumour is found and which show no history of trauma, it is better described as a group in its own right.

There are many other causes that have to be mentioned like strokes, hydrocephalus, syphilis, meningitis, brain abscess, disseminated sclerosis, Parkinsonism, syringomyelia, Paget's disease. But all of these, if they do produce smell abnormalities only produce them rarely and as a symptom so minor that it is hardly worth mentioning among the others.

TRAUMA

Although trauma will include the type of loss of smell which may follow neurosurgical procedures such as transfrontal exposures, it is to head-injuries that we refer when considering post-traumatic anosmia.

The first reference to loss of the sense of smell as a result of head-injury was in 1864 by Hughlings Jackson:

'In 1837 a Gentleman of Sheepwash in Devon was struck from his horse. All the worst effects of concussion resulted—his sense of smell was lost for ever.'

This statement implies that he was well aware of the condition, and it would be very strange indeed if the association between trauma and anosmia had not been noticed before, at least by those so afflicted.

Nevertheless, in the medical literature, at any rate, groups of two or three cases appeared, presented as though they were a rarity. It is not until 1943 that a major series was published by Leigh. Stimulated by Elsberg's work, he tested the sense of smell in 1000 head-injuries and found it to be defective in 72 cases. Leigh's patients were not entirely unselected, as they were soldiers who had been taken to a special centre with rather severe head-injuries during the second world war. Piacentini (1949) published the results of examination of 155 cases, but the most important and most carefully analysed series is that of Sumner (1946). He studied 1167 patients who had been admitted to hospital or been seen in its casualty department with head trauma no matter how trivial, and a second series of 101 patients referred with post-traumatic anosmia.

Another useful series is one of 1800 patients seen by Rowbotham (1966) and some references by Lewis (1966).

TABLE 3. *The Incidence of Anosmia in Head-injury*

Series	Anosmia in	Number of cases
Leigh	7·2%	1000
Piacentini	9·7%	155
Sumner	7·5%	1167
Rowbotham	10·5%	1800
Lewin	5·0%	—

The incidence appears to be just below 10 per cent of all patients with head-injury. As will be seen later, even extremely trivial injuries may cause anosmia, and this feature may be overlooked both by the patient and the examiner unless it is specifically looked for.

Relation between anosmia and the severity of the injury

Leigh gave no correlation between severity and anosmia, but concluded in general that the most serious injuries were the most likely to damage the sense of smell.

Sumner studied this aspect in detail, using the length of the post-traumatic amnesia as an indication of the severity of the injury. This is an acceptable guide and his results showed a marked rise in incidence of anosmia for increased length of amnesia. In the very long amnesia cases, however, there was a slight decrease in incidence, a fact which is difficult to explain.

Although there is no doubt that the more severe the injury the more likely it is to cause anosmia, the trivial injuries should not be overlooked. It is possible that a number of anosmias of unknown aetiology may be the result of very minor injuries which have been forgotten. Sumner quotes the case of a chef carrying a tray of cooked vegetables who struck his head just above the eye against some scaffolding. As soon as he took hold of himself again he realized that he could not smell the vegetables. A girl with complete anosmia was seen by the author. She had slipped backwards on some highly-polished floors striking the back of her head, but got up immediately as she did not lose consciousness. As soon as she was up she noticed that the very powerful smell of polish had disappeared.

Cases are rarely as obvious as these, as most patients do not notice their loss of smell for some time, and even then rarely report for examination at once. Apart from diagnostic reasons, the medico-legal questions which are so often present in these cases demand that olfaction is tested as soon as the patient is seen even after very minor trauma.

Relation between anosmia and the site of injury

According to Rowbotham, apart from the increased percentage of frontal injuries in his patients with anosmia, the other sites were more or less random.

Sumner calculated that out of the total number of cases where the site of injury was known, the incidence of anosmia for frontal, vertical and occipital regions was 3·9 per cent, 10·4 per cent and 21 per cent. It should be noted, however, that as the cases of frontal injury were more numerous, the absolute incidence was greater in those.

An interesting fact that emerged was that if anosmia was present, an occipital blow was five times more likely to cause anosmia.

Prospects of recovery

Out of Leigh's 72 cases only six recovered. All did so within four months except one which recovered after a year. The serious nature of Leigh's cases, however, has already been commented upon.

By and large, in unselected cases including every degree of severity, one-third will ultimately recover. In mild cases 50 per cent can be expected to recover, but if the post-traumatic amnesia has been over 24 hours then 90 per cent will have permanent anosmia.

Recovery takes place in 75 per cent of the cases within the first three months, but cases have been recorded of recovery up to five years. The rate of cases which recover is very fast during the first ten weeks; after that it is very slow and incomplete. This would imply that there are two mechanisms for recovery. The first permits early and full recovery suggesting that neural tissue has not been destroyed, the time involved being only necessary for clearing of oedema and blood clot. The second type with inadequate and late recovery indicates destruction of neural tissue. There is a very close similarity to the behaviour of lesions of the facial nerve. A number of cases develop parosmia, smells being sometimes completely altered, usually becoming unpleasant, but nevertheless dis-

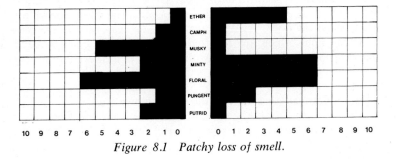

Figure 8.1   Patchy loss of smell.

tinguishable one from the other. Sometimes parosmia is simply a phase in recovery.

Those cases which do not recover adequately do not show a general uniform reduction in sensibility, but produce two types of response. The first shows a patchy loss of smell, being greater for certain smells than for others. The olfactory spectrogram demonstrates this very clearly in a case of the author's (Fig. 8.1). The second type may show a slightly reduced threshold or again a patchy appearance, the actual smell perceived for each odour being a rather sweet and sickly smell. This Single Non-Discriminating (S.N.D.) response has already been discussed in Chapter 6.

Partial anosmia

Rowbotham states that 20 per cent of anosmias were unilateral and complete, 14 per cent unilateral and partial. Bilateral anosmia

—probably more severe cases—were mostly complete. Clearly unilateral anosmias are easily missed.

Hyposmia has not been fully investigated, but the results of the case shown in (Fig. 8.2) indicate that quantitative and qualitative investigation of these cases, such as by the olfactory spectrogram, may give very interesting results.

Cause of the anosmia

The cause of anosmia following head-injury is essentially a matter for speculation. Post-mortem examinations are usually on very seriously injured patients and there is no means of knowing what their olfactory status was. There is a record of one case of posttraumatic anosmia who died years later and where a tiny spicule of bone was found embedded in the olfactory tract at post-mortem.

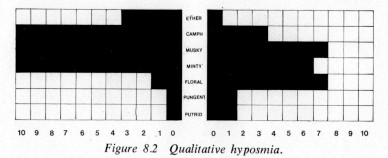

Figure 8.2 Qualitative hyposmia.

Clearly such an event is a possibility even when no fracture is visible at X-ray. Tearing of the fine olfactory nerves as they pass through the canals of the cribriform plate is especially likely when there has been a fracture through the cribriform plate with a cerebro-spinal fluid leak. When the blow has been occipital the nature of the injury to the brain is clearly that of *contrecoup* and a shearing tear of the olfactory nerves is possible. Avulsion of the olfactory tract may occur in fractures of the fronto-ethmoid, while compression of the tracts and bulbs could be due to oedema or to accumulation of blood-clot. Injury to the central connections of the olfactory fibres, or the cortical areas of olfactory perception is even more conjectural. It would be interesting to study the ability for identification of odorous substances by patients in whom temporal lobe injuries might be suspected.

The management of post-traumatic anosmia

This starts as soon as possible after the patient has regained consciousness. Particular note should be made as to the site of injury, the length of the post-traumatic amnesia and other neurological signs such as injury to other cranial nerves.

Examination of smell cannot be undertaken until the nose is clear of blood-clot or cerebro-spinal fluid. A bolster should be placed across the nose but no drops or other medication inserted. As soon as smell can be tested, this should be done and the result recorded, at least with simple aromatic substances such as coffee, lavender, peppermint. An olfactory spectrogram is of much greater value and should be done if possibl (Fig. 8.3) shows the spectrogram of a patient taken 4 hours after a severe malar injury which had left him unconscious for some minutes. He was not complaining of

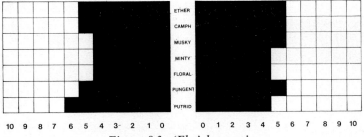

Figure 8.3   *'Flat' hyposmia.*

anosmia at that time, though he became conscious of this as a result of the test. Further spectrograms were made 10 days later and 3 months later. A series of improving results can permit a good prognosis to be given.

The prognosis soon becomes the most important question. If the olfactory spectrogram shows the existence of hyposmia rather than anosmia, then the prognosis becomes very hopeful particularly if improvement is apparent on further tests. If total anosmia is present, the patient should be warned of the possible permanency of the condition, but it should be explained to him that complete recovery is still possible, particularly if his injury was not severe. He should be seen again 3 months later and it is only at this point that a firm prognosis can be made. If recovery is apparent then there is still a good chance of further recovery but if anosmia is still present then the condition should be accepted as permanent. There are a few reported cases of recovery after up to 5 years, but they are meaningless to the individual patient who should not be misled. He should be warned of the dangers of anosmia, particularly in regarding to burning and escaping gas.

If the patient has developed parosmia, Sumner has suggested that in some cases this may precede recovery. In the author's opinion this could be accepted as evidence of recovery provided the olfactory spectrogram shows hyposmia only. On the other hand, if the parosmia is associated with a Single Non-Discriminating (S.N.D.) response, as sometimes happens, then recovery is not likely.

No treatment can give any benefit. Numerous suggestions have been made—galvanism, on the grounds that electrical stimulation of any form will encourage nerve function; vitamin B_1 which is at any rate harmless; and various 'cures' can have at least a restful effect after head-injury. That at the Mont Doré had acquired a substantial reputation on the continent among the less modest financially with olfactory disorders. What should be totally forbidden is the use of intra-nasal decongestants, sprays and inhalations which can in the long run only add to the patient's troubles. Steroids are still in vogue.

Prospects for future study

The most useful advance would be a technique to distinguish between neurapraxia of the olfactory nerve and the more permanent tissue destruction. This would apply to all cranial nerves and, although possible in motor function tests such as in the facial nerve, it does not appear imminent regarding sensory nerves.

The study of hyposmia is, on the other hand, entirely feasible using quantitative and qualitative techniques. Information regarding the selectivity of damage and recovery would be important to the understanding of sensory neuronal function. Fatiguability studies could provide useful results in those cases where there is enough smell present to test this. Testing hyposmic patients for odour identification ability may be associated with this, but care would have to be taken regarding total mental capacity.

Anosmia following nasal surgery

Anosmia has been reported following major sinus surgery as well as minor surgery of nasal septum or even polypectomy. Plastic surgery of the nose has also not been without its complaints. It is difficult to assess these reports because, as is usual in olfactory assessment, techniques of testing are diverse and often in case reports on details are given other than a recording of the incident.

The author has therefore to fall back on the cases referred to him for further opinion or for medico-legal purposes. These have fortunately been few, as there are only six among the records.

1. Four of these had had minor nasal surgery usually involving more than one procedure: submucous resection of the septum, nasal polypectomy, submucosal diathermy of the interior turbinates and intra-nasal antrostomy. In one of these had there been recorded a leak of cerebrospinal fluid and they were all seen by the author between 2 and 6 months after surgery. O.s.g. testing demonstrated the presence of smell in all of them, showing a pattern which was more typical of vasomotor or allergic rhinitis than of injury. As this was the initial illness which led to surgery it is very doubtful that an inquiry was involved at all. The value of nasal surgery to patients who have loss of smell from nasal conditions has, of course, been discussed in the preceding chapter.

2. In one case the surgeon had attempted to drain a frontal mucocoele by an intra-nasal route. At one stage in the operation a vigorous flow of cerebrospinal fluid was observed and the surgeon packed the torn area with a free muscle graft. Soon after the patient had a trans-frontal repair using fascia. He had complete and permanent anosmia, but it is not clear whether this followed the initial injury or the repair operation.

3. The last case was one where a plastic surgeon had removed a mid-line dermoid of the nose in a 16-year-old girl. The anosmia which followed was demonstrably hysterical but the background could not be clarified by the author.

In other words, except in accidents such as (2), anosmia is most unlikely to follow nasal surgery and there is no reason why the patient should be warned specifically that loss of smell is one of the hazards of nasal surgery.

TUMOURS

The types of intra-cranial tumours which may affect the sense of smell are the following:

Osteomas.
Meningiomas, especially of the olfactory groove.
Frontal lobe tumours.

Tumours and swelling around the optic chiasma.
Temporal lobe tumours.

There are two ways in which these lesions may affect the sense of smell:
1. By pressure or interruption of the olfactory nerve fibres or the olfactory bulb.
2. By interference with the intra-cerebral pathways from the bulb to the cortical areas.
It is important to make this distinction because of the different olfactory findings in these two groups. These differences have been based on the work of Elsberg and Stewart (1938), Spillane (1938) and Zilstorff-Petersen (1953, 1959). Although the author has not had the opportunity of examining as many patients as these workers, he has nevertheless had the opportunity of testing their findings in enough patients to confirm their validity.
1. Pressure on the olfactory nerves or tract such as that exerted by an olfactory groove meningioma, a supra-sellar meningioma, an extensive pituitary tumour or an aneurysm of the internal carotid artery will raise the olfactory threshold. Comparison between one side and the other is particularly valuable. Fatigue will not be prolonged.
2. In supra-tentorial intra-cerebral neoplasms and in the large dural growths which are buried in the brain tissue the olfactory threshold is normal but fatigue is prolonged on the side of the neoplasm.
3. Large frontal-lobe tumours may extend to the under-surface of the lobe and in that way exert direct pressure on the olfactory bulb and tracts as well as affect the intra-cerebral pathways. These cases show both elevation of threshold and prolongation of fatigue on the affected side.
4. Tumours affecting the anterior part of the temporal-lobe have a normal threshold and a prolonged fatigue.
5. There is a tendency for an odorous stimulus applied to the side in which the tumour is present to be felt as if it came from the opposite side. Spillane suggests that this may represent a defect in odour recognition and he refers to it as 'olfactory alloaesthesia'.
In Elsberg and Stewart's 101 supra-tentorial tumours they obtained correct localization in 74 per cent of the cases. No localization was possible in 22·9 per cent but incorrect findings were made in only 2·9 per cent. In 150 patients who were eventually found

not to have a tumour, olfactory tests had given no evidence of a localized lesion.

Among the intracranial tumours the most common are the gliomas. These are of three types:

Astrocytomas which can occur at any age, may be cerebral or cerebellar and are relatively benign;

Glioblastomas are tumours of middle life, arise in the cerebral hemisphere and are extremely malignant;

Medulloblastomas are very malignant tumours which usually arise in the cerebellum in children. Olfactory signs are, of course, not involved in these cases.

Second to the gliomas are the meningiomas, then come the acoustic neuromas, tumours of the pituitary and other rarer tumours and lesions. These include secondaries from distant tumours, sarcomas, tuberculomas, vascular malformations and cholesteatomas.

The features of the less uncommon lesions will be briefly reviewed as a differential diagnosis cannot be made without some knowledge of their behaviour.

1. Osteomas

These may arise from the inner table of the skull or in the paranasal sinuses. The latter are the more common, but they may grow to a considerable size with hardly any symptoms. Eventually they may erode into the orbit or into the cranium with visual disturbances or unilateral loss of smell appearing before other symptoms. The dangers of orbital and intracranial infections are very great in these cases because of the communication with the sinuses that has been created.

Osteomas have a characteristic appearance on X-ray from which the diagnosis can be made.

2. Meningiomas

These are slow-growing tumours which tend to occur in middle-life. They are essentially benign unless they are allowed to grow and cause irreversible damage to the brain tissue. This, together with the fact that they can be surgically excised, makes it imperative that they be diagnosed early. Meningiomas of the anterior cranial fossa, particularly if they lie in the olfactory groove, sometimes present with loss of smell. This may be total anosmia or hyposmia and, if unilateral, such an olfactory abnormality is suggestive of a meningioma. Radiography is essential in the diagnosis and may show a number of important features. There may be a localized hyperostosis or an area of intracanial calcification close to the cortical surface and also focal erosion of the calvarium.

3. Frontal-lobe tumours

When these tumours are pre-frontal, i.e. confined to the part of the frontal-lobe which lies anterior to the precentral gyrus, they may present with headache only. Papilloedema and vomiting may only appear late, if at all.

Mental symptoms are common, the patient showing a progressive dementia in which the most characteristic feature is a failure to grasp a situation as a whole although the different aspects may be understood. As the disease progresses the intellect deteriorates so that the patient becomes stupid and careless, finally becoming incontinent and showing no feelings of propriety. Convulsions occur in most cases and some show an expressive aphasia.

The grasp reflex is a well-known sign but is not always present. Usually it is found in the hand opposite to the side of the lesion, but if the tumour is in the upper part of the lobe it may be shown in the foot only.

Increasing pressure on the corticospinal fibres will cause weakness on the opposite side, particularly the face and tongue, while pressure on the olfactory nerves and tract cause a homolateral anosmia. As the tumour extends backwards it will compress the optic nerve to cause optic atrophy while the rise in intracranial pressure causes papilloedema on the opposite side. The combination of anosmia, blindness with primary optic atrophy on the same side and papilloedema on the opposite side is known as the Foster Kennedy syndrome.

If the tumour involves the pre-central region, excitation of the corticospinal fibres causes focal convulsions.

4. Tumours around the optic chiasma

These may cause pressure on the olfactory tracts and resulting loss of smell. This is associated with visual disturbance due to pressure on the chiasma resulting in changes in the field of vision as well as optic atrophy. The most frequent lesions are the following:

1. Pituitary tumours

The most common are adenomas. The eosinophil and basophil tumours are hormone-producing and it is the resulting syndromes that are primary features in these conditions. Chromophobe adenomas, although producing signs of hormone deficiency, are more likely to give symptoms of compression. A headache in the lower part of the frontal region near the mid-line is a common occurrence. It is not due to raised intracranial pressure but to the stretching of the diaphragma sellae.

The visual fields are affected early and examination of the fundus oculi shows pallor and loss of substance of the disc. Rarely papilloedema is present.

A lateral X-ray of the skull shows erosion of the anterior or posterior clinoid processes. The sella turcica is ballooned outwards and there is no evidence of calcification.

Malignant tumours of the pituitary or sphenoid sinus are not common, but they produce bony destruction which is visible on X-ray.

2. Craniopharyngiomas

These are tumours originating from the remnants of the embryonal Rathke's pouch. The tumour may appear above or below the diaphragma sellae so that the symptoms are somewhat different. If it is above, there is visual disturbance and hypothalamic compression leading to hydrocephalus. It is in these cases that olfactory symptoms may be present while those patients with the tumour below the diaphragma show signs of pituitary dysfunction.

3. Suprasellar meningiomas

These rare meningiomas of middle life cause optic atrophy and bitemporal defects of the visual field. If the tumour is small and does not involve the pituitary or brain tissue it could be removable.

4. Aneurysms

Aneurysms can occur in the circle of Willis or any artery in the region around the pituitary.

5. Suprasellar cholesteatoma

These cysts are rare and have similar presentation to the meningiomas except that they tend to occur in younger people.

Taken as a group, whenever loss of smell is associated with changes in the field of vision and optic atrophy, a tumour of the region of the optic chiasma must be suspected. Radiography is carried out as described in Chapter 6 and is an essential part of the investigation. Tumours within the sella turcica may balloon it out and there may be erosion of the anterior or posterior clinoids. A suprasellar meningioma may show a characteristic thickening of the bone in the anterior clinoid region. Aneurysms often show calicification in their walls while malignant tumours produce bone destruction. The craniopharyngioma shows an enlarged sella and, characteristically, areas of calcification, so that a picture of a swollen sella with calcification within it strongly suggests a craniopharyngioma.

5. Temporal-lobe tumours

If the tumour is on the right, focal symptoms may be only slight. Taste and smell are usually impaired but the patient is not totally anosmic in lesions of the uncus. Uncinate fits with olfactory or gustatory aura are often present in these anterior tumours. There may be bilateral impairment of hearing together with tinnitus and auditory hallucinations. Left-sided lesions may cause aphasia and speech may be very disorganized. Important localizing signs are the visual field defects which occur in almost half the patients. This is due to the involvement of the lower fibres of the optic radiation which pass round the inferior cornu of the lateral ventricle and causes a crossed upper quadrantic hemianopia with a more extensive loss on the ipsilateral side. The olfactory aspects of temporal-lobe epilepsy are discussed in the following section.

EPILEPSY

Brain's text-book on diseases of the nervous system begins its definition thus: 'Epilepsy is a paroxysmal and transitory disturbance of the function of the brain which develops suddenly, ceases spontaneously and exhibits a conscious tendency to recurrence.'

This disease has long been known in history and although recorded as far back as the Laws of Hammurabi it is to Hippocrates that the first attempt to place it in a 'scientific' context is attributed. Nevertheless, belief in the divine or demoniacal origin of the 'sacred disease' has persisted into the present century.

Epilepsy can be considered a symptom of disturbed cerebral function. There are, of course, many other causes of disturbed brain function such as those due to abnormalities of the cerebral blood-flow, brain damage by injury or tumours, metabolic abnormalities and mental illness.

Epilepsy is associated with abnormal, usually synchronous electrical discharges recordable from the brain and the disturbance of brain function results in disturbance of movement, sensation, behaviour or consciousness.

A clinical classification which has been used for a long time is the division of epilepsy into 'primary' or idiopathic and 'secondary' or symptomatic. The primary groups those cases which appear to have a constitutional basis. This group is getting smaller as more cases can be placed into the second.

The secondary group includes those cases of epilepsy caused by intra-cranial lesions such as trauma, infection and tumour, or extra-cranial causes such as anoxia, toxins and poisons.

Unfortunately, this classification, though useful in separating those cases where the underlying pathology is known from those where it is not, is insufficient to describe the nature of the symptoms that occur. For our purpose it may be of value to discuss it as referring to seizures of Major, Minor and Focal type.

1. MAJOR SEIZURES

These are also known as 'grand mal'. They are characterized by a prodromal period of anxiety and tension which may last from hours to days. This is followed by a sudden loss of consciousness, rigidity, apnoea. Often the patient may urinate. Tonic spasms of the whole body including the face takes place, followed by clonic movement or jerks of the arm, legs, face and body. Gradually these wear off and the patient is left comatose, gradually regaining consciousness. Rarely this fit is preceded by an aura or warning phenomenon.

2. MINOR SEIZURES

These are known as 'petit mal' because they consist only of temporary absences or lapses—a transient break in the continuity of consciousness which lasts a few seconds.

3. FOCAL SEIZURES

Ever since Hughlings Jackson described fits caused by lesions of the uncus in 1888 it has been recognized that epilepsy can have a focal origin in the brain. It is also clear that these cases fall into the 'Secondary Epilepsy' group and that there must be an underlying pathology. Focal seizures produce motor, sensory or psychic disturbances of cortical function and may constitute the whole syndrome. Sometimes the focal symptoms will precede a generalized seizure in which case they are usually referred to as 'the aura'. It should be clear, however, that this so-called aura is itself the focal seizure and therefore an indication of secondary epilepsy suggesting some underlying pathology.

As the symptoms which occur depend on the site of the lesion, they have considerable diagnostic value. Seizures can originate in:

(i) The motor cortex causing clonic or tonic movements of a limb, or part of a limb.

(ii) The sensory parietal cortex when paraesthesiae or dysaesthesiae may be experienced.

(iii) Occipital cortex producing visual abnormalities. These seizures may involve a temporary loss of vision or visual hallucinations such as flashes. Occasionally bizarre seizures occur with hallucinations of detailed and complex scenes, but in those cases it is likely that the temporal-lobe is also involved.

(iv) Frontal-lobe seizures may have no focal manifestations but occasionally there are psychic elements.

(v) Temporal-lobe. This is the most common form of epilepsy as it comprises one-third of the cases and it is in this group that abnormalities of smell are generally found. For this reason temporal-lobe epilepsy will be considered in detail.

For the purpose of description anatomists have called 'temporal-lobe' the part of the brain that lies below the sylvian fissure and blends posteriorly with the occipital lobe (Fig. 8.4). Its connections, however, form a unity including the hippocampal gyrus, uncus, hippocampus, the amygdala, cingulate gyrus and associated areas. This whole is known as the limbic system.

In this system is found the higher cortical representation of the autonomic system including the reproductive system. It influences the reticular formation and the hypothalamus and through that the pituitary gland and has been known as the 'visceral brain', responsible for integrating our autonomic systems and our emotions. The type of epilepsy being considered can arise anywhere in the limbic system and its most characteristic form is the psychomotor attack.

These focal temporal-lobe seizures may be single or complex and may or may not be followed by a generalized seizure such as a period of amnesia. The symptoms which are experienced during these seizures are very varied and include epigastric sensations often associated with oral sensations. There are memory disturbances

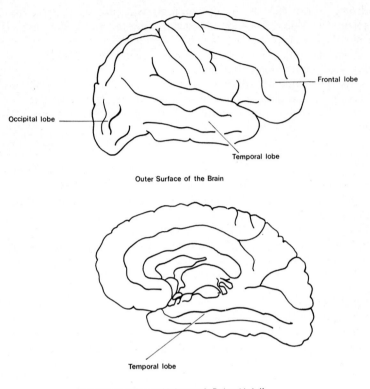

Outer Surface of the Brain

Temporal lobe

Inner Surface of Temporal Lobe seen in Brain cut in half

Figure 8.4   Temporal lobe.

such as the *déja vu* or *jamais vu* phenomenon and feelings of unreality or depersonalization. Primary automatism may occur and
there may be disturbance of emotion such as sensations of fear,
anxiety or depression.

Finally there are hallucinations of smell, taste, hearing, vision
and movement as in vestibular epilepsy. Is is to the olfactory symptoms that we shall be referring in more detail.

The causes of temporal-lobe epilepsy are multiple (Falconer,
1970). Among the 'gross' lesions are cerebral tumours, arteriovenous
malformations, depressed fractures, gun-shot wounds, and scars
from intra-cranial abscess. As these lesions account for only a small

proportion of the cases, the majority were thought to be idiopathic but, as a result of correlation of post-mortem findings and the inspection of temporal-lobes excised at operation, the pathological substrate to the condition is better understood. This approach has been greatly enhanced by the gradual accumulation of electro-encephalographic findings and their careful comparison with cinical data (Margerison and Corsellis, 1966). The fact that temporal lobectomy is now being carried out in intractable cases of epilepsy has also permitted detailed histological examination of the area. The pathological findings in a series of 100 consecutive patients submitted to operations in which the hippocampus, amygdala and uncus were in the resected specimen were reported by Falconer, Serafetinides and Corsellis (1964). Very similar findings in another hundred patients were later reported by Falconer and others (1968).

We have, therefore, some considerable knowledge of the lesions present in temporal-lobe epilepsy:

Mesial temporal sclerosis

This has also been known as Ammon's horn sclerosis but in fact the sclerotic process usually involves the amygdala and the uncus as well as the hippocampus. This is likely to be an acquired lesion and not an inherited one, but in Falconer's series there was a positive family history in 15 per cent of the cases so that there must be a genetic factor as well. It is probably the result of an asphyxial episode in infancy such as febrile convulsions, and the damage caused to the tissues of the limbic system at a time when they are particularly vulnerable causes them later to sclerose producing an epileptogenic focus. It is unilateral in 90 per cent of cases and is responsible for over half the patients with temporal-lobe epilepsy.

Hamartomas

These are benign congenital malformations with a strong genetic element. Although they are usually abnormalities of the glia, there are occasional patients with small capillary angiomas, dermoid cysts or areas of tuberose sclerosis. In Falconer's series these lesions accounted for 20 per cent of the cases.

Miscellaneous lesions

These represented less than 20 per cent in the same series as

F

scars and local infarcts. Some of these scars were secondary to mastoid disease, but it is not clear whether the majority of these scars were the cause of the epilepsy or the result of injuries obtained during seizures.

The remainder of Falconer's series showed equivocal or non-specific lesions which are not uncommonly found in brain tissue not associated with epilepsy.

Association between smell and the epileptic seizure

Jackson in 1888 first noted the association of seizures with an olfactory aura and lesions of the uncus; but even earlier, Gowers (1881) pointed out that an epileptic crisis could be stopped by a strong stimulus. He suggested that in the case of an olfactory aura the seizure could be stopped by inhaling a powerful and unpleasant odour.

Since those early times, the anatomical, neurophysiological and clinical relationships of temporal-lobe epilepsy have been carefully studied and it is possible to consider the olfactory phenomena in three parts:

The olfactory aura;

The role of olfaction in the facilitation or inhibition of an epileptic seizure;

The disturbances in the sense of smell experienced by epileptics.

(a) THE OLFACTORY AURA

Epileptic seizures which begin with an olfactory aura are rare. In Gibbs' and Penfield's series they accounted for only 2 to 3 per cent of the cases.

The olfactory hallucination is usually unpleasant and is either defined as 'organic' in type, usually as a smell of putrefaction, decay or faeces, or as 'chemical' like petrol, chloroform or ether. It may be a smell of burning, or an indefinable sensation. Only occasionally it proves agreeable like perfume.

The smell usually appears to come from outside, but occasionally the patient feels that the smell comes from within himself.

Sometimes the olfactory sensation represents the whole seizure and there are no other phenomena, but usually it is associated with other features:

(i) An emotional element. Some attacks may consist solely of the olfactory hallucination followed by inexplicable anger—sometimes uncontrollable rage. There may be a feeling of anxiety or intense fear. A powerful sense of imminent death can be experienced, or the patient may experience profound sadness and tenderness. Occasionally there is a feeling of pleasure and joy which may lead to a state of beatitude.

(ii) Memory changes. A panoramic vision of the past is not uncommon and this vision of past events may be extraordinarily precise and detailed. There may be a sense of familiarity with these events—the well-known *déja vu, déja fait, déja entendu* phenomena or a sense of total strangeness as *jamais vu*, etc.

(iii) Motor changes. Movements of the nostrils may take place automatically. They may be simple dilations of the nares or more complex movements which produce grimaces. A sensation of hunger may be accompanied by mastication, salivation, tongue movements or even suction. Occasionally a putrid smell is followed by nausea and even vomiting.

(iv) Other sensory phenomena. It is not uncommon to find the olfactory hallucination accompanied or preceded by others which may be auditory, visual or sensations of movement.

An association with variable gustatory sensations is very common and probably occurs in over half the cases.

A few cases where olfactory hallucinations were followed by pleasurable sexual sensations have been reported, but although referred to the genitals, they do not appear to produce orgasm.

Passouant (1965) suggested that the physio-pathology of the 'uncinate crises' should be considered in two parts, as the clinical manifestations are in two parts. In the first instance there is the olfactory hallucination and in the second a feeling of depersonalization or 'dreamy state'.

The olfactory sensation is suppressed after section of the secondary fibres to the contralateral hippocampus or after excision of the uncus. In other words the olfactory aura can be related to the area concerned with normal olfactory perception. The electrical potentials induced in this area, from the pre-pyriform cortex to the uncus and connected to the olfactory bulb and receptor cells, by olfactory stimuli have a short latency.

The psychosensory phenomena occur when the 'non-olfactory' rhinencephalon discharges. This area includes the dentate gyrus and Ammon's horn.

(*b*) OLFACTION AND THE FACILITATION OR AN EPILEPTIC SEIZURE

There is a well-known type of epileptic fit which can be induced by sensory stimuli. The most common is the seizure set off by intermittent light. It appears that there is a lowering of the convulsion-threshold associated with a sensory-inducing factor.

It is rare, however, that an olfactory stimulus has been shown to set off an epileptic fit. Passouant quotes Ionaseau's work with such reflex rhino-epilepsies. The latter was apparently able to induce clinical or electroencephalographic seizures using strong-smelling substances in about 35 per cent of the patients he tested. These had temporal-lobe epilepsy with different types of symptoms and he used a variety of substances including chloroform, ether, ammonia, xylol, benzene, lavender, peppermint and aniseed. These were repeatedly offered every 20 or 30 seconds. There was no olfactory aura in these cases except the more common auditory and visual phenomena, so that the stimuli had no direct olfactory meaning but apparently acted in a non-specific manner.

Passouant believes that the olfactory stimuli have a general facilitating effect on the cerebral structures. Like all sensory perceptions, they alter the degree of alertness which depends on the intensity of the stimulus as well as other environmental factors. These changes are probably the result of activity in the reticular formation and are accompanied by electrical changes in the hippocampus.

Patients who have had electrodes placed deep in the brain prior to neuro-surgery were tested with a number of smells: coumarine, cineol, exaltolide, and methyl-nionone. Each produced electrical activity in a parahippocampic focus which was enhanced by odour recognition. Trigeminal nerve stimulation by non-odorous chemical and mechanical stimuli also produced electrical changes and these were enhanced by a simultaneous olfactory stimulus. This suggests an important interaction between the olfactory sense and common nasal sensation as far as their affects on the structures of the limbic system are concerned.

As far back as 1881, Gowers noted that some seizures could be arrested by strong sensory stimuli such as ammonia or amyl nitrite. Efrom (1956) studied this inhibitory effect on an uncinate crisis with e.e.g. control. He managed to interrupt the progression of the seizure from the dreamy state towards a generalized episode by disagreeable odours such as hydrogen sulphide or strong odours such as concentrated jasmine, whereas pleasant smells such as essence of lavender, 'Bergamot Messina', rose, etc. were unsuccessful.

In 1957 Efrom reported an extraordinary experiment. He tried to induce a conditioned reflex to a non-specific stimulus in the same way as a dog can be made to salivate by ringing a bell. He used the patient who had learnt to arrest a seizure by smelling jasmine and then associated this smell with the sight of a silver bracelet. A full second order conditioned reflex developed so that the patient could interrupt the seizure by looking at the bracelet.

(c) DISTURBANCES OF SMELL

Little work has been done regarding the permanent changes in olfaction experienced by epileptics. Certain patients have reported changes in taste and smell for a greater or lesser length of time before a seizure. Sometimes this can persist for days. Similar changes can occur after a seizure, occasionally producing a complete anosmia, but more commonly a parosmia. These do not usually persist for so long.

The author has studied a small group of nine patients who experienced olfactory changes and they fell into three groups:

Parosmia

Four patients. Three of these experienced unpleasant smells and taste which were difficult to describe but which were revolting in nature. Foodstuffs produced these disagreeable flavours even though the patient was aware that they had been normally prepared. One patient experienced vague and faint sensations which he was unable to describe but which were not unpleasant.

Hyposmia

Three patients. All of these had a raised threshold on o.s.g. but no anosmic zones. In one patient this hyposmia was unilateral but all three also had poor gustatory discrimination and an electro-gustometric threshold which was above average.

Hyperosmia

Two patients. These were the most bizarre because they both experienced normal smells with exceptional intensity, one patient feeling unable to eat. A common feature was a very powerful emotional element which, on questioning, appeared to be associated with abnormal feats of memory.

Passouant has also noted olfactory changes in these patients and suggests that the eletrical discharge involves not only the limbic system but also the olfactory bulb and even the receptor cell. The olfactory changes can represent a form of post-critical exhaustion of the olfactory pathways.

Santorelli and Marotta (1965) measured the olfactory threshold of 90 epileptics in a psychiatric hospital using six substances: phenyl ethyl alcohol, musk, citral, guaïacol, vanilla, pyridine. They showed a general lowering of the threshold as compared with members of the staff, but after a spontaneous or an induced seizure they invariably showed a raising of the threshold which took more than twelve hours to revert. Unfortunately the authors of this interesting work do not give us any details as to the nature of the epilepsies.

Management

Some cases which present with olfactory symptoms can be very difficult to recognize, often because the patient does not consider the associated symptoms relevant. The author is very familiar with the similar situation of vestibular epilepsy. Often the patient who complains of vertigo will withhold the story of his 'dreamy state' and similar feelings until his condition is gone into in great detail. So that when temporal-lobe epilepsy is suspected, the relevant features in the history must be elicited, and the investigations, described in Chapter 6 (e.e.g., X-rays, etc.), carried out.

It is important in this way to separate those suffering from 'primary' epilepsy from those where it is 'secondary' to a demonstrable underlying pathology.

In the treatment it is important for the patient to understand the nature of his problem and its implications. Therapy is with anticonvulsant drugs.

Ever since the use of bromides started in the middle of the last century an increasing number of anticonvulsant drugs have been produced. They fall into a number of categories:

The barbiturates, such as phenobarbitone and primidone. These are used usually in combination with phenytoin to control grand mal seizures and are probably best avoided in the type of temporal–lobe seizures we are mainly concerned with here.

The oxazolidinediones, like troxidone and aloxidone, are used in petit mal and should not be used in these cases either.

The succinimide derivatives. The same applies to this group although methsuximide (Celontin) can sometimes help in controlling secondary epilepsy in combination with phenytoin.

The hydantoins. This group usually provides the drug of choice.

In most of the author's patients phenytoin (Epanutin; Dilatin) was used. Occasionally another drug such as sulthiame has to be added.

The dosage of phenytoin used was either 100 mg b.d. or 100 mg t.d.s. It has a few minor side effects which should be recognized, such as gastric symptoms, gum hypertrophy and occasionally sensitivity reactions. Hirsutism has been reported. When taken over a period of years it may cause a megaloblastic anaemia, but blood dyscrasias are rare. Lymphadenopathy is an unusual side-effect which regresses very rapidly after the drug is discontinued. If the serum concentration of phenytoin rises above 20 μg per ml, nystagmus occurs and this is quite common. If it is allowed to continue to rise to between 30 and 40 μg per ml ataxia will develop. This cerebellar dysfunction disappears quickly, however, as soon as the dosage is reduced.

Methoin, another hydantoin, is more toxic than phenytoin, and ethotoin, although of low toxicity, is not very potent.

Carbamazepine (Tegretol) is a dibenzapine derivative and therefore quite different type of drug. It is extensively used in trigeminal neuralgia but also has anticonvulsant properties and controls focal cortical seizures. It is not the drug of choice but can be added to phenytoin in resistant patients.

It is wise to build up the dosage until control is achieved while looking out for toxic side-effects. When the seizures have been well controlled at this dosage for some months, it can be again reduced to a minimum effective dose. If the drug of choice used by itself is not entirely effective, a second drug may be added in the same manner.

Neurosurgery has an important place in the treatment of secondary epilepsy. Lesions such as meningioma or abscess require surgery. Post-traumatic scars can sometimes be excised if the symptoms are very disturbing and in the same way porencephalic cysts may also be excised. Temporal-lobe epilepsy which cannot be controlled by drugs may require a temporal lobectomy.

OTHER ORGANIC CAUSES

Other types of intracranial disease have been reported as causing loss of smell. It should be stressed, however, that smell abnormalities in these cases are neither as common nor as important a part of the symptomatology as in the three main groups just described:

trauma, tumours and epilepsy. The author, whose practice consists of patients referred specifically for the assessment of olfactory changes, has little personal experience of them.

According to Barraquer Fevrè and Arumi Fargas (1952), abnormalities of smell can be found in cerebral abscess, vascular lesions of the frontal lobe, meningitis, disseminated sclerosis, syphilis, syringomyelia, Paget's disease, although on rare occasions.

Klippel and Lohermitte (1909) described in detail the 'nasal crises' in tabes which included parosmia as well as hypersecretion of nasal mucus. Smell changes have also been reported in other from of syphilis such as general paralysis of the insane.

Hydrocephalus can cause olfactory abnormalities because of the pressure on the relevant structures of the brain.

Alvarez (1966) studied in great detail what he called 'little strokes' and the abnormalities of smell which may accompany them. These strokes may be brief and silent and are rarely correctly diagnosed because patients do not report sudden attacks of dizziness, vomiting, falling spells or blackouts. The patient may deteriorate, becoming dull or forgetful, irritable and unreasonable. Very often the patient is quite unconscious of these big changes in his life and will complain of the smaller, even trivial, symptoms. It is here the smell abnormality may cause concern. Loss of smell is a fairly common symptom and usually appears suddenly, commonly associated with bad taste. It can be unilateral and may be associated with other sensory loss, such as hearing. Others (Fisher, 1958) complain of smelling unpleasant odours either emanating from themselves or from the environment.

PSYCHOGENIC DISORDERS

Reference to bad smells exists in the Biblical literature as well as in the medical commentaries of ancient Egypt. These reports are too brief to evaluate but since then history has continued to provide comments on olfactory symptoms which may well have been the product of disordered minds. Louis XI is a well-known case, with a complex and disagreeable personality associating the constant belief that there was a bad smell about him with suspicion of all the individuals who remained around him. There is no doubt that his delusions had an important effect on the history of France and that it may have been a valuable rather than a damaging one. The writers reporting these historical facts owe their survival more to the poetry than to the science of their works and it is therefore never clear whether reports of a good or bad smell are not simply expressions of emotion rather than of fact. Furthermore, the ordinary problems of translation take on dangerous forms when discussing the sensory system. French, for instance, does not contain a word for 'loudness'. Other forms like 'higher' or 'stronger' are used instead and the exact meaning provided by the context. In scientific descriptions this is not sufficient and might be totally misleading. The single word *sentir* has to be used when 'feeling' or 'smelling' is intended. Thus *se sentir mal* means 'to feel unwell' and *sentir mauvais* means 'to smell bad'. Apart from its psychological and linguistic relevance, this has a more direct impact on the study of olfaction as a large proportion of the reported cases are to be found in the French journals. The difficulty of translating from the French has been fully understood by Pryse-Phillips (1968). He gives the most thoroughly documented review of the literature on olfactory hallucinations in both languages as well as important papers in others. These reported facts were subjected to a minute statistical and descriptive analysis. He reports a personal series of 137 cases in which all the clinical variables are analysed by means of a computer.

The classification of symptoms which emerged from this study has already been described with the abnormalities of smell in Chapter 6, but it is worth recalling briefly here:

(1) Illusions of smell
   (i) alteration of smells
   (ii) alteration of smells with emotional value
(2) Hallucinations of smell
   (i) synaesthesiae
   (ii) true hallucinations
   (iii) pseudo-hallucinations
   (iv) hypnosis
(3) Abnormal sense-memory

The abnormal sense-memory, although an event of particular psychological importance, can hardly be described as a disorder unless it is associated with the *déja vu* phenomenon or similar perceptual changes in temporal-lobe epilepsy. Its mechanism is far from clear, indeed, even its place among the vast range of changes in the human mind is difficult to define. The name of Marcel Proust will always be connected with it, because that strange and most important of novelists describes how he dipped his *madelaine* or biscuit in his coffee and brought it up to his mouth. The aroma which stimulated his olfactory sense immediately conjured up a simultaneous collection of associated memories based on a loving and protected childhood. From this event there stretched thirteen volumes, significantly entitled *à la Rechèrche du Temps Perdu*, and which have added something of great value to the whole of succeeding literature. Perhaps Proust used this incident simply as a clever introduction to a novel, but the profound role which smells can have in evoking memory cannot be denied. It may be that what is evoked is not memories in the chronological sense but an emotion. It is the state of mind of one's childhood or past which is suddenly and temporarily regenerated with all its beliefs, fears and lack of experience of fore-knowledge. The impression can only be described as the simultaneous re-emergence of all the past in the form of memories. The implication of such an incident other than the personal or literary ones is very great as it is possible to imagine the presence of the feelings of the past as well as the events in a physiological sense in the brain. They can be brought into consciousness by an olfactory stimulus and may well be recordable as electrical potentials.

Olfactory illusions and hallucinations are much more common. They can be associated with organic brain lesions such as tumours and epilepsy as has been described, as well as alcoholism and drugs such as L.S.D. In that form they can be part of a general hallucinatory syndrome. They can also be found in patients with psychiatric disorders. Alliez and Noseda (1945) found that among 2000 male patients in mental hospitals, 13 had hallucinations of smell or taste, and among 3400 females there were 82. Expressed as percentages, it comes to 0·65 and 2·4 for males and females respectively. The smells they complained of were nearly always unpleasant and were very varied in type; they altered frequently and sometimes disappeared without any improvement in the general condition of the patient. Usually the olfactory symptoms are somewhat imprecise but occasionally they can be very clearly defined. Alliez and Noseda split up their patients into groups with diagnoses of *délires polymorphes dégénératifs*, persecutory delusional states and dementias, which terms are not commonly used in the English literature, but summarized by Pryse-Phillips they can be broken down as follows:

Schizophrenia-like psychoses 82 per cent
Confusional states 6·4 per cent
Alcoholism ⎫ less than
Senile dementia ⎬ 5 per cent
Manic-depressive illness ⎭ each

Other writers have shown markedly different figures and it should be remembered that this series comes from a French mental hospital at the end of the Second World War. The selection of in-patients was very specific of that place and period and olfactory hallucination is probably much more common in the non-schizophrenic illnesses.

For the French writers, olfactory symptoms could be found scattered in two groups of cases:

1. The hallucinatory psychoses which could be acute or chronic. And these were considered in their widest sense.

2. The others, in which were lumped together the depressive illnesses, alcoholism, confusional states.

The psychoses can then be subdivided further into those where olfactory hallucinations are lost amidst a rich and divers symptomatology, and those where the smell is the dominant symptom. It is to these that Alliez gave the name *délire à base olfactif*. These cases are the rarest and the prognosis is better than where there is a multiplicity of symptoms.

The diagnosis and management of mental illness cannot properly be discussed here, but a summary of the type of psychiatric problem that should be considered when patients complain of smells that are not apparent to others may be found of value.

1. Schizophrenic disorders

This is a group of illnesses which are both serious and very common and show a hereditary predisposition. The majority of the beds in mental hospitals are occupied by these patients and the onset of the disease is commonly between the ages of 15 and 25. Often there are traits in the personality which antedate the illness, but environmental factors appear to have a precipitating influence. The patient may show signs of disorder of thought which may vary between difficulties in concentration or association of ideas and severe breakdown into total incomprehensibility. This is usually associated with an emotional impoverishment as well as lack of energy or initiative. Delusions are common, sometimes referring to break-up of thought processes, often hypochondriacal in nature and very bizarre. Hallucinations of taste and smell become built in to the delusional system and frequently there are auditory hallucinations. Willis and Banister (1965) based their diagnosis of schizophrenia on 10 or 15 symptoms.

2. The Olfactory Reference Syndrome (O.R.S)

This name was coined by Pryse-Phillips to represent a group of patients who showed no evidence of schizophrenia. They are shy, sensitive, self-contained and restrained. Usually young men under 30 years of age, the olfactory symptoms, which are true hallucinations, are the dominant feature of the illness. There are attempts to mask or wash away the smell, together with the social withdrawal which shame and anxiety would engender. This produces the development of ideas of reference and a misinterpretation of the environment. The smells are unpleasant and appear to originate from the body. These patients show no evidence of schizophrenia but produce an ashamed, self-critical and contrite reaction.

A typical case was that of one young man who, although highly strung, was socially active. He suddenly perceived that he had a bad body smell which persisted for months. He became anxious and unhappy, giving up much of his social life and misinterpreting signs and gestures as though people were complaining about him. He used deodorants, and washed excessively. Five months' psychiatric treatment failed to remove the adour.

3. Depressive illness

As a symptom, depression is probably the most common complaint that patients make to their doctors. It not only varies in degree, but can occur in virtually all the psychiatric illnesses as well as in many physical illnesses.

The symptomatology of depressive illnesses includes loss of interest, lack of energy, poor appetite and a disturbance of sleep which typically results in early wakening between 2 a.m. and 4 a.m. The feelings of depression show a diurnal variation often with ideas of self-reproach, guilt and despair. It is probably from these ideas that delusions develop in severe depression. These delusions may involve belief in the patient's own guilt for imagined crimes, or that they have an incurable disease, often 'shameful' in origin. Occasionally these delusions are accompanied by hallucinations which may be auditory or olfactory.

Depressive illness tends towards spontaneous remission, but apart from the anguish experienced there is always a risk of suicide which makes recognition of this disease imperative.

Since the introduction of anti-depressant drugs, the treatment of many cases of depressive illness has been greatly facilitated. They can often be effectively and safely treated by the general practitioner or ear, nose and throat surgeon. It is therefore quite reasonable, if a patient with olfactory symptoms presents a general picture of depressive illness, to treat him with antidepressant drugs for a trial period of four weeks. There are two main types of drugs:

1. Tricyclic antidepressants

These are related to imipramine. They usually act after about a week and their effect can only be properly assessed after three or four weeks and they should only be reduced gradually in case of recurrence.

Trimipramine, for instance, should be given in a dose of 25 mg tds increasing to 50 mg tds after a week. Treatment is usually continued for several months. Other drugs are amitriptyline, imipramine, etc.

2. Monoamine Oxidase Inhibitors (M.A.O.I.s)

This is a mixed group varying in dosage and effects. They should not be taken with cheese or wine or Marmite, Bovril, yoghurt or alcohol, as sudden rise in blood-pressure may occur. These drugs include iproniazid, isocarboxazid, phenelzine sulphate, tranylcypromine.

If, after four weeks, there is no improvement or if there are suicidal tendencies the patient should be referred to a psychiatrist. Tricyclic and M.A.O.I.s together are best left to the psychiatrist to use, and other modes of treatment such as psychotherapy and E.C.T. may also be necessary.

4. Epilepsy

This has already been described but it is mentioned here again because it can easily be mistaken for, or indeed associated with, psychiatric illness.

5. Hysteria and malingering

These are dealt with in some detail in the chapter on the legal aspects of smell.

CHAPTER NINE

The Legal Aspects

There are three different ways in which the sense of smell becomes involved in legal processes. If it has been impaired as a result of injury there may be litigation; it may be the subject of accident insurance; it may be associated with pollution of the atmosphere.

I. LOSS OF SMELL THROUGH INJURY

Loss, diminution or alteration of smell which may result from an injury can lead to benefit under the National Insurance Acts, to a case for damages brought in the civil courts or a claim under a privately arranged insurance policy.

1. The national insurance benefits

The statutory scheme of insurance against industrial injuries is embodied in the National Insurance (Industrial Injuries) Act, 1965, which itself was designed to consolidate the Acts of 1946 to 1964 and certain related enactments and supplementary regulations made by the ministries administering National Insurance.

Those insured include all persons employed in Great Britain under any contract of service or apprenticeship, whether written or oral and whether expressed or implied. This also includes those working on board British ships or aircraft or craft registered in Britain or owned by a person resident in Britain. It excludes those employed in the service of their own husband or wife or even near relatives if they reside in the same house.

The source of funds necessary to pay out benefits is collected as weekly contributions from both employer and employee.

Benefit is of three types:

Injury benefit is payable to a person for the period in which he may be incapable of work.

Disablement benefit which is by way of a disablement gratuity or

pension and is payable to the insured person if he suffers a loss of physical or mental faculty as a result of the injury.
Death benefit—payable to relatives.

Loss of the sense of smell falls into the second category of disablement benefit. Benefit is not payable in respect of an accident happening while the insured person is outside Great Britain except if he is a mariner, an airman, or working on the Continental Shelf. For the purposes of the Act, an accident arising in the course of an insured person's employment shall be deemed, in the absence of evidence to the contrary, also to have arisen out of that employment. A point which is not always understood is that an accident is deemed to arise out of an insured person's employment even if that person had been acting in contravention of any regulation or against any orders given by his employer. Since 1961 this regulation has been extended to include those injured by the misconduct of other persons, by an animal and even if he is struck by lightning.

If the sense of smell is lost as a result of such an injury, then disablement benefit can be claimed. This should be done within a period of 156 days (Sundays being disregarded) from the accident, but will not be available until the third day of that period or after return to work. There are a number of circumstances in which this benefit can be increased such as the existence of dependants, cases of special hardship or of unemployability. This situation may arise if the insured person was a perfumer by profession or a specialist in some type of work in which the retaining of a good sense of smell was an important requisite. Disablement benefit may be given out in the form of a gratuity if it is small or of a pension if it is larger.

The determination of claims and questions is divided between the Minister and a medical board or medical appeals tribunal. The questions to be determined by the Minister are of a general sort and include whether a person is or was in insurable employment; whether he or his employer were exempt from payment of contributions, etc. Questions concerning the disablement such as whether the relevant accident has resulted in a loss of faculty, or at what degree this disablement should be assessed must be referred to and determined by a medical board or a medical appeals tribunal. The medical board consists of two or more medical practitioners of whom one is the chairman, and in the appeal tribunal there are two medical practitioners and a chairman; all these members are appointed by the Minister.

The claimant is referred by the insurance officer to a medical board for determination of the disablement questions. If he is dissatisfied with the board's decision he may appeal and his case will be referred to a medical appeal tribunal. If fresh evidence becomes available, such as non-disclosure or misrepresentation by the claimant or any other person of a material fact, the original decision may be reviewed by a medical board at any time. If there has been an unforeseen aggravation of the disablement the assessment can also be reviewed.

There may be a further appeal allowed by a medical appeal tribunal on the grounds that the decision is erroneous in a point of law. This appeal may be made to an Industrial Injuries Commissioner within three months of an assessment at the instance of the claimant or an association of employed persons of which the claimant is a member or the Minister.

Insured persons are also covered against 'prescribed disease', even though these may not be caused by an accident, provided such a disease or injury is due to the nature of the employment. Included among these is poisoning by fumes and vapours of the following compounds which may damage the olfactory area.

Lead, manganese, phosphorus or phosphine, arsenic, mercury, carbon bisulphide, benzene or a homologue, nitro-, amino-, chloro-derivatives of benzene or a homologue or nitro-chloro-benzene. Dinitrophenol or a homologue or by substituted dinitrophenols or by the salts of such substances.

Tetrachlorethan.
Tri-cresyl phosphate.
Tri-phenyl phosphate.
Diethylene dioxide (dioxan).
Methyl bromide.
Chlorinated naphthalene.
Nickel carbonyl gas.
Nitrous fumes.
Gonisma kamani (African boxwood).
Chrome ulceration and inflammation or ulceration of the skin or respiratory passages or mouth resulting from the use or handling of chromic acid, chromate, or bichromate of ammonia, potassium, sodium, zinc or any preparation.
Beryllium or a compound of beryllium.

Exposure to cadmium fumes.
Inflammation or ulceration of the mucous membrane of the upper respiratory passages or mouth produced by dust, liquid or vapour.

2. Damages

Apart from disablement benefit to which any insured person is entitled, the claimant who has lost his sense of smell can sue for damages in the civil courts. This is usually done under the tort of Negligence or, if it is a work accident, under a breach of Statutory Duty such as an infringement of the Building Acts or other infringement of safety requirements. In the case of an injury sustained at work, both may be claimed, whereas in a road traffic accident, for instance, it would be a question of negligence.

These cases may be tried in the county courts if the claims for damages are less than £750 or in the High Court if the sum is greater than that. As the claims for damages for loss of smell are usually for considerably more than that figure the cases are tried in the High Court. Witnesses to the accident will be called and statements from the doctors who may have examined the claimant heard, and medical experts may be questioned. The judge will then give damages as a global figure composed of two types: special damages compensating any expenses incurred by the claimant including loss of wages, and general damages relating to the loss of a faculty. Half the value of the disablement benefit which the claimant may have been granted is deducted from the special damages.

One hundred years ago damages of this type were given by juries rather than judges and were so erratic that they are not worth considering when calculating reasonable expectation of damages. A number of more recent judgments are given here as precedents and they will be discussed subsequently:

Batten v. *Harland & Wolff Ltd.* (1961) 1 Lloyd's Rep. 261; March 2, 1961; Salmon J. The plaintiff was a man aged 56 at the time of trial. He was a riveter employed by the defendants. He fractured his skull but made a remarkable recovery, helped by his own courage and tenacity. He was well enough to go back to his full work after nine weeks. He suffered from fairly frequent headaches and occasional giddiness, but there was a strong probability that these would clear up entirely. The only permanent effect was loss of his sense of smell. He did not mind working on a ladder up to ten feet high. He did on occasions before his accident work

35 feet above ground, but he was no longer called upon to do this because he was now somewhat nervous of working as high as that. This factor might militate against him if he was on the labour market, but he was a good workman and had been with his present employers for 40 years, and it was extremely unlikely that he would be dismissed before his working life was over. There was no diminution in his wage rate. Salmon J. assessed the general damages that would have been recoverable on the basis of 100 per cent liability at about £558.

Phillips v. *Leggett* 1968 C. A. no. 96; March 5, 1968 Danckwerts, Diplock and Sachs L. JJ. Appeal from O'Connor J. The plaintiff suffered a fractured skull. He was in hospital for four days and resumed work within six weeks. During the two-and-a-half years after the accident he had suffered from headaches, vertigo, loss of libido and a certain amount of irritation. Within about a year the headaches had become very infrequent and very transient. These symptoms were likely to clear up. He had permanently lost his sense of smell and a greater part of his sense of taste, although he could distinguish between sweet and sour and salt and the like.

The Court of Appeal dismissed the defendants' appeal against O'Connor J.'s award of £3250 general damages. *Per* Diplock L. J. The award was at the high end of the bracket and probably higher than most judges would have awarded.

Kearns v. *Higgs & Hill Ltd*. (1968) 112 S. J. 252; March 13, 1968 Danckwerts, Diplock and Sachs L. JJ. Appeal from Rees J. The plaintiff, who was aged 27, was employed in the building industry. His nose was broken causing a partial obstruction of the nose and a complete loss of his sense of smell. He returned to work after three weeks but was able to do only light work and was put on the job of tea-boy. He had slight dizziness for a time. He had had headaches for the four-and-a-half years since the accident, but these were likely to improve rapidly once the action was over. Some slight improvement in the nasal obstruction was expected as time went on. The loss of sense of smell would be permanent, but he still had his sense of taste. He was unable to smoke and drink in the quantity which he used to before. His nose had mended well and his appearance was described by the medical experts as being not uncomely. However, his nose was not the same shape as before and the plaintiff did not like it. There was apparently no continuing loss of earnings.

The Court of Appeal increased Rees J.'s award of £850 general damages to £1500. *Per* Diplock L. J.: £1300 to £1400 was a proper figure for the loss of sense of smell and associated disabilities of the nose. To that must be added £100 to £200 for the headaches for four-and-a-half years.

Thackray v. *Henry Willcock & Co. Ltd.* 1968 C. A. No. 150; April 4, 1968 Willmer, Diplock and Edmund Davies L. JJ. Appeal from Faulks J., Birmingham Assizes. The plaintiff was a man aged 52. At the time of the accident he was the site clerk on a building site. He was struck on the head by a baulk of timber and sustained a fractured skull and concussion. He had loss of memory for 10 days. He was detained in hospital for a fortnight. He had permanent anosmia in the right nostril but retained at least a partial sense of smell in the left nostril. His sense of taste was normal even for the more sophisticated flavours, which indicated that his sense of smell had substantially recovered. The chance of his developing epilepsy was now trivial. He was fit for all forms of work except possibly climbing on scaffolds. He genuinely believed that he still suffered from a number of other serious disabilities and had worked for only short periods during the three-and-a-half years after the accident. The court accepted the medical evidence that he had made a virtually complete recovery and was fit to return to work within six months and that his residual complaints were of a psycho-neurotic origin and had resulted from his failure to take reasonable steps to rehabilitate himself.

The Court of Appeal reduced Faulks J.'s ward of £1750 general damages to £1000.

Lewis v. *North London Group Hospital Management Committee* (May 26, 1966; Thesiger J.). Fitter, aged 61. Concussion. Fractured skull. Laceration of scalp. Loss of all sense of smell. Considerable degree of loss of sense of taste. No improvement expected. Occasional headaches expected to clear up when case over. Returned to light work a few weeks after accident. Still on light duty. *General damages*: £1500: (1966) C.L.Y. 3324.

Bromley v. *Pringle* (*Policy*, March 1967; January 13; Browne J.). Male, aged 29. Lorry driver. Injured in road accident. Damage to nerve in nose causing total loss of sense of taste and smell. No loss of earning capacity. *General damages*: £1350.

Leese v. *National Coal Board* (*Policy*, June 1968; Leeds Assizes, Brandon J.). Miner, aged 55. Injured in colliery accident in August 1964. Loss of sense of smell and taste for which no organic cause. Pain for over two months. Senses of smell and taste would probably return within six months. Off work about 17 months. Judge found should have returned to work within 10 months. *Damages*: £950.

Ludlow v. *Lawes Chemical Co.* (*The Times*, March 14, 1967; Stephenson J.). Male, aged 49, who worked as a fitter and welder. Face splashed by sulphuric acid. Total loss of sense of smell. Had been unable to smell escaping gas when wind blew out flame on gas-stove at home. Partial loss of sense of taste. Could distinguish only between bitter and sweet and recognized only highly flavoured food. Judge accepted that injury was of a functional nature resulting from shock of accident and that recovery not impossible. *General damages*: £900.

Storey v. *Jewson Bros.* (January 16, 1957; Havers J.). Male, aged 48. Painter earning £10 5s. a week. Fractured skull and loss of sense of smell. *Award*: £3000.

Penfold v. *Brown* (April 21, 1964; Elwes J.). Male, aged 40. Fractured skull. In hospital for four weeks and then out-patient for three months. Away from work for three months but able to work regularly after that. Although he had been unconscious for two hours and partially insensitive for five days, he was lucky by comparison with others. Suffered certain amount of giddiness but remarkably free from pain. Loss of senses of smell and taste permanent and 'pretty complete'. Had twice been unaware of escaping gas. Pleasure at table seriously impaired. Considerable deprivation to be left merely to feed to keep body and soul together. *Agreed special damages*: £224 14s. 4d. *General damages*: £2250.

Johnson v. *Wastie* (November 2, 1954; Hilbery J.). Male, aged 40. Farmer and grain merchant. Fractured skull; sense of smell lost; sense of taste almost destroyed; sight and hearing affected; headaches. *Award*: £2000.

Counsel v. *Cough* (June 11, 1954; Monmouthshire Assizes, Lynskey J.). Male, aged 38. Omnibus driver. Lost all sense of taste and smell. *Per* Lynskey J.; 'This is a loss of one of the delights of life.' *Award*: £1592 (including special damages).

Cornell v. *Barnett* (May 2, 1956; Donovan J.). Female, aged 21. Shorthand typist. Fractured skull causing permanent blunting of senses of taste and smell. *Award*: £1500.

Armatowski v. *Leach* (May 3, 1961; Diplock J.). Male, aged 41. Metal grinder earning £10 a week. As result of severe assault, suffered broken nose at bridge, black eye and lacerations. Off work for ten days. Was now deprived of sense of smell and finer sense of taste. Cannot now distinguish sweet, bitter, sour and salt. Can no longer enjoy country walks. No loss of earning capacity. *Award*: £1200.

Brooks v. *Graham* (October 8, 1963; Winn J.). Male, aged 43. Jamaican, store assistant earning £12 a week. Concussion and shock. Eight days in hospital. Only permanent disability was loss of sense of smell and greatly impaired sense of flavour. *General damages*: £475. *Special damages*: £54.

Bell v. *London Transport Executive* (April 16, 1956; Lloyd-Jacob J.). Female, aged 52. Housewife. Fractured skull. Nerve damaged which resulted in permanent loss of sense of taste. *Award*: £940.

Almeroth v. *Evelyn Bros.* (October 5, 1955; Ormerod J.). Female, in fifties. Housewife. Fractured skull and collar-bone. Loss of smell probably permanent. *Award*: £750.

French v. *W. J. Barton* (December 21, 1956; Ashworth J.). Male, aged 70. Retired haulage contractor. Burns to mouth. Loss of sense of taste. Could no longer enjoy food, beer and cigarettes. *Award*: £500.

Wagman v. *Vare Motors* (October 16, 1958; Thesiger J.). Male, aged 51. Master furrier. Nose broken resulting in deformity. Partial loss of smell and taste and sinusitis. Also had hernia and post-traumatic neurosis as a result of which had headaches and depression. *Award*: £375.

Reading the reports of some of these cases a few features become quickly apparent. They are not entirely comparable to each other in the extent of injury resulting in loss of smell, yet they are similar in many ways and reading report after report involves considerable repetition. Despite this there appears to be a great variation in the damages granted ranging, in this group, from Wagman *v.* Vare Motors (£375) to Phillips *v.* Leggett (£3250) although only ten years between them. An appeal against this high award was dismissed. Another case in the same year (Kearns *v.* Higgs and Hill Ltd. 1968) had produced general damages of £850 but on appeal this was raised to £1500. In that case Lord Justice Diplock stated that £1300 to £1400 was a proper figure for the loss of the sense of smell and associated disabilities of the nose.

When claims are made for loss of smell and taste the relationship between these two senses must be made clear by the medical experts otherwise the malingerer or the claimant suffering from an associated neurosis is likely to obtain more money than the well-balanced claimant who expresses his symptoms with clarity. This subject will be discussed further with the examination of the patient.

EXAMINATION OF THE PATIENT

Careful notes must be made at every point if medico-legal implications are apparent and all cases of injury of any sort will automatically be considered from that aspect from the moment they are seen by a doctor. This may often be in a hospital casualty department where notes should be very detailed. They can be produced in court as doctors are not expected to remember details of an examination, but these notes should have been made at the time and not altered subsequently. They should include accurate observations and a detailed physical examination as well as X-ray findings. If certain tests such as X-rays cannot be done, then the reason must be recorded, e.g. the patient was not fit enough for the procedure to be carried out.

Anything that the patient says about his physical or mental state is admissible evidence provided it is relevant and not just hearsay but it is for the court to decide on this question. Any witnesses are also interviewed and their stories recorded. In all cases of head injury a complete general examination is carried out with particular emphasis on the nervous system and the cranial nerves. All the senses are tested specifically and, in the case of smell, this must be done as soon as the patient can co-operate. There are so many cases where anosmia is only noted weeks later and this will lead to some doubt when it comes to court. Any bruising, laceration, bleeding from the nose or a nasal leak of cerebro-spinal fluid is relevant to this question. X-rays must always be taken if this is possible.

The medical records made under the National Health Service can be used only by authority of the Medical Officer of Health and Regional Board, and solicitors have no legal right to see them. They can ask for a report, however, which requires permission from the patient. Fees cannot be claimed under the National Health Service (paragraph 14 of the terms of service) unless a special examination has been requested.

If a report is required on the loss of taste and smell in a personal injury case, a history is first taken in an attempt to recognize the type of injury. Those which are known to cause loss of smell have been fully discussed in Chapter 8. The present complaints are then noted. These may be of complete or partial anosmia or parosmia and complete or partial loss of taste. It should be clear in the examiner's mind what is meant here by smell and taste. Smell is the sense which allows perception of odoriferous molecules. The end organ is in the roof of the nose and the impulses it produces are carried to the brain by the first (olfactory) cranial nerve. Taste is a totally different sense involving the stimulation of the end organs (taste buds) of the surface of the tongue by chemical substances. These substances produce only four 'tastes': salt, sweet, sour and bitter. The impulses from the anterior two-thirds of the tongue are carried by the chorda tympani, a branch of the seventh (facial) cranial nerve. Those from the posterior third are carried by the ninth (glossopharyngeal) cranial nerve. It is clear that these four lingual modalities cannot possibly account for the innumerable 'tastes' which we perceive and that smell is responsible for these 'flavours'. Unless the patient is a good observer or has some degree of sophistication he will generally be unable to distinguish clearly between these senses, and if he has lost the sense of smell he will usually complain that he cannot taste also. In law this point should be clear and in the assessment of the patient it is important. When a plaintiff claims that he has lost the ability to smell and to taste even substances which are salt, sour, sweet or bitter the suspicions of the examiner should be aroused. This is because it is anatomically unlikely for all these nerves (olfactory, facial and glosso-pharyngeal) on both sides to be injured simultaneously unless the injury is so severe that the patient is not expected to survive with only such minor defects. Before suggesting malingering or hysteria the surgeon should check that the patient could taste before the injury. Persons who are heavy smokers may have very little in the way of taste buds left on the tongue and persons who have had severe ear infections or operations on the ears may have had the taste nerve, the chorda tympani, damaged in the process (Arnold, 1973). If a plaintiff states that he is unable to smell or taste but on testing is able to distinguish salt and sweet with his tongue, he is likely to be genuine as what he says can easily be explained by an injury simply involving the olfactory mechanism. It is perhaps better for the purpose of description to refer to three types of perception:

Loss of:		*Mediated by*:
Smell	1. Smell ⎫	Olfactory nerve
Taste ⎰	2. Flavour ⎬	
⎱	3. Taste	Chorda tympani and glossopharyngeal nerve

In the case of damage by fumes, this anatomical question does not apply as they could, of course, damage not only the olfactory area but also the tongue.

The surgeon should then question the patient directly about any previous illness which may have caused loss of smell. These are usually nasal abnormalities such as polyps, chronic sinusitis and allergic rhinitis.

Clinical examination of a plaintiff who has lost the sense of smell includes inspection of the face and head for scars and signs of injury with particular reference to the nose. The latter is examined with speculum and head-mirror and a note made for any abnormality which may be relevant. A septal deviation caused by an injury is relevant if the patient is complaining also of nasal obstruction. Such a deviation however is never the cause of anosmia and a pre-existing malformation cannot therefore be relevant. Nasal polyps, allergic rhinitis, etc. frequently affect the sense of smell so that their presence in the nose may be important. The postnasal space, throat and ears are routinely examined and all the cranial nerves tested. There is no need in a case of this type to examine the other body systems or to carry out a general neurological examination unless a specific reason demands it.

The tests for smell and taste are of necessity entirely subjective and the examiner must rely on the truthfulness of the plaintiff who claims that he can or cannot smell or identify a substance. Because of this, certain techniques become necessary in testing these individuals. Tests for smell and taste have been described in detail in Chapter 6, as well as the responses given in hysteria and malingering. In practice, if a plaintiff complains of complete anosmia there is no point in attempting to carry out an olfactory spectrogram. The different odorous substances should be placed under his nose and he is asked to sniff and say whether he can smell anything at all and, if he can, to try and recognize. If there is some olfaction present, then olfactory spectrogram should be carried out to produce some quantitative and qualitative pattern of response. The plaintiff who claim total anosmia, particularly if he claims total loss of taste as well, may be suspect and requires further testing. The unexpected introduction of a foul smelling substance like skatole may cause a reaction meaningful to the experienced eye and the response to ammonia is also of value. A genuinely anosmic individual will either say he can smell it because he may mistake the sensation of burning (transmitted by the fifth cranial nerve) with that of smell or he may well say very precisely that he cannot smell but he can 'feel' it. Tricks of this type, although distasteful, are necessary in these circumstances but even they are not entirely foolproof so that the examining surgeon, when giving an opinion, should explain the manner in which he has reached it. The sense of taste is also tested as described in Chapter 6 and of course the integrity of the fifth (trigeminal) nerve and its nasal branch also tested with the other cranial nerves.

After such an examination has been carried out with the plaintiff's permission a report is prepared. This includes the date of the examination and report and the place where it was carried out as well as the name and age of the plaintiff and the date of the injury. A summary of the facts is given as they have emerged from the plaintiff's own story and any other reports and investigations which may be available. Any relevant facts in the past medical history are then given and a description of the present condition including the examination. A summary of all this leading to a diagnosis forms the conclusion and this is followed by a prognosis. As this is often a most important feature, it is necessary for the examiner to be well acquainted with the known facts relating to head injury and anosmia (Chapter 8). Non-medical matters concerning earnings, for example, should not be included. A fee for such an examination and report can be claimed and varies, of course, with each individual surgeon. In most cases it will be the surgeon's usual fee for a consultation plus an extra sum for the report. If the surgeon is then called upon to act as a witness before the High Court, a much larger fee can be claimed as well as expenses.

By far the most common cause of claims both for disablement benefit and for damages arising from an injury with olfactory sequelae results from head-injuries. These are mainly road-traffic accidents or work injuries. Much less common is loss of smell resulting from noxious fumes or liquids, and there is a different type of case where the plaintiff claims damages from the surgeon for injury to the olfactory area during surgery. These cases are rare and often difficult to prove. Some are clearly due to negligence as, for instance, the case where a surgeon aiming to pack the nose with a gauze strip soaked in anaesthetic solution used the wrong liquid, burning the nasal lining with trichloracetic acid. The author has been asked to see a few patients who claimed loss of smell after a nasal operation (one was for excision of a septal dermoid by a plastic surgeon) and in all of these hysteria or malingering were proved. That does not mean that the olfactory area cannot be damaged during nasal surgery; on the contrary, intervention near to any sense organ such as smell or hearing can damage that organ irreparably despite all precautions, and it should be recognized as an operative risk. In the case of olfaction that risk is so small that it is not necessary to bring it specifically to the attention of the patient before each nasal operation. It should be classed with very rare self-evident risks which indeed include death, and which recognize that surgery may fail and may make matters worse. The important thing from the legal point of view is to make sure that all due precautions and care have been taken. In the United Kingdom all medical practitioners should be insured against litigation as the costs can be extremely high.

The aim in an examination for legal purposes is to decide whether a plaintiff is truthful in his complaint, as well as the degree of loss of faculty which he has suffered. When truthfulness is not in question, the medical report and subsequent legal action is a matter of routine. It is the case where the plaintiff's veracity is in doubt which is not only a diagnostic problem for the surgeon and a legal one for the judge but an important social and psychological one. It is the frequency with which this situation arises which makes it an important question. The plaintiff may be a 'malingerer' which means that he has formed a calculated and fraudulent attempt to deceive with the aim of gaining an advantage to himself, either financial or the evasion of work, or both. There are a few such individuals, but any surgeon with these cases would agree that they are by far in the minority. Yet there are many indeed who exaggerate their symptoms when compensation is involved. When the symptoms become more than exaggeration and form an entity of their own, the physician has often been perplexed even as to how to name the condition. It is often called hysteria or, because it is always associated with compensation of some form or another, it has been known as 'compensationitis', 'litigation neurosis', as well as 'accident neurosis' and 'post-traumatic neurosis'.

Like any other clinical syndrome, it will have many grades of severity but there will always be a multiplicity of symptoms accompanying the main one which may be of loss of smell.

Accident neurosis or post-traumatic neurosis

The most important work on this subject has been produced by Miller (1961, 1966) based on a very large clinical experience and research. He drew attention to this very common problem which had previously been generally neglected in the literature by analysing over 4000 patients examined over a period of 12 years for medico-legal purposes. Since then the problems involved have become widely known and frequently commented on.

The development of psycho-neurotic complaints is twice as common after industrial as after road accidents. It is more than twice as common in men as in women. Although it is rare in children, age appears to have no bearing on the development of such symptoms. If analysed against the severity of the injury there is no doubt that there is an inverse relationship to severity. It is much higher, in fact, among patients who have not been unconscious than among those who have. A number of predisposing factors become apparent; thus the incidence of accident neurosis is related to social status and in the industrial group it is the unskilled or semi-skilled who develop gross neurotic sequelae. In Miller's group more than half have come from the Registrar-General's social classes 4 and 5, when only 38 per cent of the cases reviewed belonged to this class. Among patients above the level of under-foreman or charge-hand, only 18 per cent developed neurotic symptoms or prolonged incapacity. This social differential is even greater after road traffic accidents where neurosis is rare among the professional or managerial patients. Although the nature of the accident does not seem important, as the most persistent sequelae follow the most trivial blows to the head, the nature of the employment appears to be relevant. People working in large organizations or nationalized industries are much more liable than those from small businesses or farms. It is often suggested that there is a particular personality, dependent, insecure, craving sympathy and yet showing paranoid tendencies, which predisposes to this type of neurosis; but it is very difficult to assess retrospectively a person's pre-accident personality.

In most of these patients there is clear evidence of an emotional illness which must be distinguished from the post-concussional syndrome of headache, postural vertigo, irritability, faliure of concentration and phonophobia. Occasionally depressive illness follows head-injury and olfactory symptoms are sometimes exhibited (Chapter 8). These patients should not be overlooked, as psychiatric treatment is beneficial.

The prognosis is generally held to be good once the case is settled one way or another, but frequently the symptoms persist in those who had a previously neurotic personality.

The legal aspects present numerous problems but Miller's view regarding whether they should be considered as a psychiatric illness is certainly challenging: 'In my considered opinion these cases are much closer to malingering than to any form of mental illness genuinely outside the patient's control, and in this context at any rate I personally find the distinction between hysteria and malingering both impossible in clinical practice and meaningless in law—a view which is shared by most lawyers experienced in this field.'

The question is whether an accident neurosis can reasonably be regarded in law as directly resulting from the accident. When a judge has to decide he often invokes two legal principles:

(i) The question of remoteness. In other words the damages claimed as the direct result of the accident must be in respect of the 'natural and probable consequences' of the occurrence. If they are too indirect or too remote they must be disallowed.

(ii) The directness of the relationship between occurrence and sequel may be interrupted by *novus actus interveniens*. Again, if an injury has led to the sequelae complained of as a result of negligent surgery or lack of proper treatment, then it cannot be held responsible.

The case for granting compensation is that it is impossible to maintain what the plaintiff would have been like without the accident, and that if he had a predisposition to neurosis then this has been tacitly accepted as a risk by the firm that hired him or by the bus company that sold him a ticket. Furthermore, the courts nowadays accept that mental suffering can be as real as physical.

The case against is that such a sequel of an injury can be regarded as a natural and probable consequence in only a minority of persons who have a predisposition. The most important objection is the subjective nature of the whole case, as the plaintiff himself is the main witness. To quote Miller again: 'Whether exaggeration and simulation are "conscious" or "unconscious", their only purpose is to make the observer believe that the disability is greater than it really is. To compensate a man financially because he is stated to be deceiving himself as well as trying to deceive others is strange equity and stranger logic.'

In general the law favours the plaintiff and most cases are settled out of court.

PRIVATE INSURANCE POLICIES

Although a small point, this is worth mentioning. Many people take out accident insurance policies and among the degrees of disability loss of the sense of smell ranks very low, usually at less than 10 per cent. In most cases this is adequate, but the writer has among his patients a perfumer who depended on olfaction for his livelihood and who lost the sense of smell completely following an injury. His insurance policy only provided for 10 per cent disability for loss of that sense. It would be wise, therefore, to advise those who depend so much on olfaction to take out a specific policy to cover it.

II. PROTECTION AGAINST UNPLEASANT SMELLS

As far as history can take us men have raised strong objection to unpleasant smells. In the 8th century B.C. when the XXIVth Dynasty of Egypt was crumbling and the Cushite pharaoh, Piankhi, marching northwards from Nubia was taking over decaying city after decaying city, the scribe wrote: 'Days passed, and Hermopolis was foul to the nose, without perfume.' Laws against fouling the air with bad smells have been enacted for hundreds of years and even in England the legal and administrative system of control over environmental pollution is by no means new. Clean air legislation existed in the early fourteenth century when a man was hanged under it for burning sea coal in London and thus making foul smoke. This severe legislation, needless to say, fell into disuse.

There are a number of ways in which a citizen may have recourse to justice to protect himself and his environment from pollution by smells.

(i) The tort of nuisance. He can sue his neighbour in the Civil Courts if he creates a nuisance by burning and making unpleasant smells in other ways.

(ii) Leaseholds and freeholds. Many leases specifically forbid 'noxious, noisome or offensive trades' being carried out in the property. This is usually enough to protect residential areas and clauses designed to prevent damage to the amenities of estate often apply to freehold properties as well. This type of protection works, of course, only on a smaller scale.

(iii) Local authority by-laws. These also give some protection against pollution of various sorts including that producing repulsive smells. Local legislation, however, has never been adequate even when reinforced by the cases brought by private citizens in the Civil Courts and this has led at various intervals to Parliamentary legislation.

(iv) Acts of Parliament. The modern system of clean air law already dates from the first Alkali Act of 1863, although this had been preceded by 15 years by water pollution legislation. In both these fields the system of control has been continuously developed.

Until October, 1969, responsibilities for the control of environmental pollution in the United Kingdom were distributed among ten different Ministers, but since then the responsibility for co-ordinating government action on this problem has been given to a single Minister. Direct Ministerial responsibility in Scotland and Wales, however, rests with the respective Secretaries of State.

In February, 1970, a standing Royal Commission was set up with the following terms of reference:

'To advise on matters, both national and international, concerning the pollution of the environment; on the adequacy of research in this field; and the future possibilities of danger to the environment.'

Over the coming years the Royal Commission will be the main source of independent advice to the Government.

The penalties for breaching any of the numerous laws which govern the pollution of the environment are written into those laws; they have therefore grown up over the years in a piecemeal fashion as the laws were introduced or amended. The British system of law in this as in related fields does not traditionally rely on the very heavy penalty as the main deterrent. It relies rather on persuasion and the belief that, especially to industrial firms, it is the disgrace that counts and not the fine. The weapon of prosecution has in the past been sparingly used. Today, however, the Government believe that the present penalties are both incoherent and generally too low. They stem from at least twenty Acts of Parliament dating back to 1804. Typical summary penalties for emitting dark smoke from a factory chimney: £100; for installing a new furnace without notice to the local authority: £20; but according to a White Paper published in 1970 the Government will examine the scale of penalties and will in due course amend the laws where necessary to bring them into line with the realities of modern life.

The control of pollution of the air by matter falls into four main parts. All of these are discussed in the 1970 White Paper on the Protection of the Environment (The Fight Against Pollution).

(a) POLLUTION BY DOMESTIC SMOKE

Domestic coal fires are still the worst source of smoke in this country and when it comes out of the chimneys of ordinary houses it hangs about at low level and so people breathe it more easily. There have been Clean Air Acts in 1956 and 1968 under which local authorities have been given power to declare 'smoke control areas' and the 1968 Act also empowers the Ministers to require local authorities to declare them. In such areas not only is it illegal to create smoke but it is illegal to sell unauthorized fuels.

In order to help people in smoke control areas to convert their open grates to smokeless heating arrangements, the local authority may offer a grant of 70 per cent of which the Exchequer bear four-sevenths. One-quarter of the houses in this country are covered by this control and there has been a rise in public expenditure for this purpose, from £2·2m in 1964 to 1965 to £5·1m in 1969 to 1970 on the smoke-control grant. The Government has not intensified its drive for smokeless areas as there has been a tight situation in the supply of solid smokeless fuels.

(b) INDUSTRIAL POLLUTION UNDER LOCAL AUTHORITY CONTROL

This covers all industry not under central Government control. The Public Health Acts and Clean Air Acts give the local authorities power to proceed in Courts against the emission of grit, dust, and dark smoke from industrial and trade premises; and authorities also have control over the height of chimneys when new furnaces are built. In practice, many authorities prefer to use their power of prosecution as a long-stop to campaigns of persuasion backed by their powers of entry and inspection. The widespread introduction of mechanical stoking for solid-fuel-fired boilers has led to a large reduction of dark smoke, and by far the greater part of industrial smoke now arises during the short periods of start-up and shut-down. Local authority control of this type covers more than 30,000 industrial premises.

(c) INDUSTRIAL POLLUTION UNDER CENTRAL GOVERNMENT CONTROL

The 1600 local councils are not able to deploy the highly skilled staff which is necessary to control industries which have an advanced technology. These industries, which carry out 'scheduled processes' in some 2200 registered premises, are mainly those concerned with electricity generation, cement, ceramics, petroleum and petro-chemicals, other chemicals, and iron and steels. They are controlled by the Alkali Inspectorate.

The 1970 White Paper envisaged the scheduling of processes in primary aluminium works, acrylic works, di-isocyanates works, mineral processing and some process not now scheduled in the petroleum industry.

Now that smoke is coming under more and more effective control, concern is being expressed over the emission of sulphur dioxide. Because of a vigilant watch kept on this problem the sulphur dioxide emission has steadily decreased since the period 1963 to 1965 when it reached $6\frac{1}{2}$ million tons a year. The Clean Air Council expects this trend to continue and estimated two years ago the following emissions over the next fifteen years:

1970	...	...	5.9 million tons
1975	...	...	5.54 million tons
1985	...	...	5.16 million tons

At low level the trend has been even more favourable; thus over the past decade the average concentration of sulphur dioxide in towns has fallen by 33 per cent.

(*d*) POLLUTION BY MOTOR VEHICLE EXHAUSTS

Control of this type of pollution is extremely complicated because of the number of substances involved, so that if the permitted level of one ingredient is lowered this will entail a proportionate increase in others.

There have been regulations against the emission of 'avoidable smoke' and smoke likely to 'cause danger' for many years and there is a British Standard (BS/AU 141) which is now being met by most new diesel-engined vehicles and which permits less smoke emission than heavy vehicles elsewhere in Europe. This standard is likely to become obligatory.

Among the substances emitted by cars are hydrocarbons from unburnt and partly burnt fuel. The evidence is that these do not harm health, but they certainly smell. These hydrocarbon emissions can be reduced by 25 to 30 per cent by fitting a simple device to the engine which will feed them back into the air intake to the cylinders.

The United States Government has announced series of very stringent standards to be met in 1973 and 1975. These are designed to reduce the emissions of hydrocarbons, carbon monoxide, and oxides of nitrogen to less than 20 per cent of their 1969 levels. They look towards completely redesigned engines which could run on lead-free petrol in the more distant future, but even at the present time their standards are higher than in Britain so that British export models have to be fitted with special after-burning equipment adding about £40 to the price of a car.

The European Economic Commission also has a standard which member countries may adopt to control emissions from petrol-engined vehicles. It does not require a greater reduction of hydrocarbon than 25 to 30 per cent but includes a reduction of carbon monoxide by about 10 per cent.

It is quite clear that the Governments of Britain and of the European Countries take a less forceful position than that of the United States. The excuse given is that the European climate is different from that of Los Angeles, and the 1970 White Paper makes a number of statements demonstrating that attitude: 'There is in fact no evidence that the carbon monoxide in our streets has any adverse effects on health or environment' and '... the development of a completely pollution-free car might not be the most sensible use of resources.' There is an obvious difference of opinion here comparable to 'how dirty is dirty?' But the tendency is likely to be towards cleaner and cleaner air whatever economic considerations restrain governments. There is a considerable amount of research going on towards replacing a petrol-driven engine by an external combustion engine, by gas-turbine and by electricity, but until any of these becomes a practical possibility motor vehicle exhaust will require constant watch.

Many techniques for the avoidance of smell and other pollutants are constantly proposed and a few types are described at the end of Chapter 11, but most nations have recognized that international co-operation is necessary as these problems are growing rapidly in scale.

INTERNATIONAL CO-OPERATION

This falls into five broad groups:

(i) Co-ordination of research and sharing of results.

(ii) Discussion, standardization and adoption of the terms in which various types of pollution may usefully be measured.

(iii) Preparing declarations commending good aims and good attitudes to governments and peoples.

G

(iv) Discussion and adoption of agreed standards and procedures for combating pollution.

(v) The enforcement of such agreements.

At present most of the international work done falls into the first three groups and this will culminate in the United Nations Conference on the Environment, in Stockholm in 1972. Until the last two groups can be applied any real progress is likely to be national rather than international.

With his increase in numbers and his advances in technology, man's impact on the environment has increased beyond all prediction and the quality of our civilization will depend to a large extent on the control we can exert on pollution. In order to do this we must extend our knowledge of biological systems and their response to pollutants, we will need to advance our technology so as to avoid pollution as much as possible and we will require a growing legal and administrative framework to encourage and enforce the necessary decisions. Most important, perhaps, will be a social, philosophical and economic analysis which will help us to recognize the priorities which make the quality of our lives.

CHAPTER TEN

Smell and Communication–
The Pheromones

Social organization in the animal kingdom depends on communication between individuals and on their appraisal of the environment. All the special senses take part in achieving this awareness but the relative importance of each sense varies with each species. Birds, for instance, are said to rely very little on olfaction and depend mainly on sight and sound, while insects and the majority of mammals are dependent on olfaction. That there is a close connection between olfaction and human behaviour has always been made obvious by the emotions aroused when smell-associations are mentioned.

The fears that suggestions of personal body odours create in Western societies are well known; the beliefs relating smell to racial prejudice, if less openly mentioned, are equally well known; the association between olfaction and sex has also been long suspected. All these questions have concerned mankind for a long time and a great deal of personal observation, deduction and thought have gone into the views expressed. Although these should more properly take their place with folklore than with scientific hypotheses, their volume and the persistence with which they continue to be discussed justify a brief review in this chapter.

In the last ten years, however, substances have been discovered and in some cases analysed and even synthesized, which influence behaviour in animals through olfaction. These substances—pheromones—may be of great importance and what is known about them will be studied in greater detail.

SMELL AND PERSONAL CLEANLINESS

Unpleasant smells caused by filth and poor ventilation have gradually been eliminated in advancing societies and it is natural that any tendency to slip back to less fastidious standards is considered a sign of the worst personal characteristics: laziness, carelessness, etc. The mere fact of being able to tolerate a bad smell in one's house is an important social sign of personal degradation. This attitude is easy to understand as representing a conditioned aspect of cleanliness, but what is more difficult to explain is the present obsession with body odours.

The smell of the body consists of numerous elements. First, there is the smell specific to the species, that is, which groups humanity as a whole; then there is the distinctive smell of the individual, and it is that smell which allows a dog to recognize its master. This individual odour is itself composite : there must be an intrinsic congenital, probably inherited, smell; there is the smell of a person's trade which clings to the garments or which is being inhaled and exhaled; the food and drink are well-known modifiers—the smell of garlic, for instance, is exuded in the sweat; illness may impart a distinctive odour, such as the sweet, ketotic breath of the diabetic. Emotions have long been suspected of altering body odour and the smell of fear can certainly induce a dog to attack while horsemen are convinced that this frightened smell can make a horse un-manageable. There is also a smell differentiating young from old and male from female. Havelock Ellis asserted that castration caused loss of the male masculine smell.

Although all these statements are conjecture and in many cases hearsay, there is much circumstantial evidence based on observation and the recent discovery of pheromones would tend to support them. This means that body odour is an important if subconscious method of communication between individuals and therefore of maintaining the social fabric. If this is the case, then the obsession with suppressing body odour by artificial means is difficult to understand. The implication is that either these modern fears will result in serious social damage by interference with human instincts, or else that it is essential to suppress these instincts if we are to live in a 'civilized' manner.

SMELL AND PREJUDICE

Although certain popular figures have been said to have excep-tionally pleasant smells—Walt Whitman was reputed to be par-ticularly fragrant—and saints were said to emit a distinct 'odour of sanctity', most allegations of unusual smell in people are of a con-temptuous nature.

Apart from individual bodily odour most groups of people have been accused by others of having a distinct and unpleasant smell. Until recently, when racial prejudice has been exposed by open confrontation, racial odours were generally assumed by the public to exist, and Negroes were also believed to have a better olfactory sense. This was, of course, held to offer further proof of their more primitive nature.

Darwin himself commented in *The Descent of Man* that 'the colour of the skin and the odour emitted by it, are likewise in some manner connected'. There is, of course, no reason why differences in pigmentation do not imply differences of other types, including smell, in the skin. Furthermore, white people are hairier than the Chinese and since sebaceous glands are so closely associated with the hair follicles they probably smell more strongly too. It has indeed long been a matter of curiosity to Europeans and Americans that the Chinese claimed a distinctly unpleasant smell emanated from Westerners; yet Bedichek (1960) tells the story of a group of Chinese girls returning home after a year in the United States. As soon as they had boarded the ship they insisted on eating only Chinese food, explaining that it would take some days before they would lose their 'Western smell' as they had learnt to their cost on a previous occasion. There is no doubt that 'racial odour' must be at least in a major part caused by diet. Everyone has experienced the smell of onion and garlic eaters, and it is also probable that meat-eaters have a stronger smell than vegetarians, although many spices produce their own odours.

All this does not exclude the possibility of genetic differences producing an underlying 'group smell', but before this can be considered a number of conditions would have to be fulfilled:

1. It must be independent of environmental factors. This condition is complicated in its meaning because groups may respond *differently* to the *same* environmental changes such as temperature, diet, etc. as a result of different metabolic pathways. One should modify this therefore by saying that groups should respond consistently and differently.

2. It must be independent of sex. Here again there could theoretically be sex-linked differences which could produce subgroups.

3. It should be independent of age.

4. It is unaltered over a long time.

5. It is inherited.

Identical twins have the same odour, suggesting that there must be a large genetic element which influences metabolism in such a way that the resulting odoriferous components of the body are altered quantitatively or qualitatively.

If racial body odour were transmitted by a single gene then according to Mendelian inheritance a hybrid would either carry that odour or not according to whether it was dominant or recessive. Children of such hybrids would then divide themselves in a 3 : 1 ratio. Furthermore, if there is indeed an association with skin colour, which feature does not have complete dominance and which is carried by more than one gene; it becomes impossible to imagine the degree of odour blending that would have to occur.

There is no doubt that if environmental differences are removed the human nose is unable to detect the innumerable odorous blends which humanity's genetic pool can produce. Until the finer chemical differences of the body's volatile emanations can be recorded no differences will be perceived, and even then such recording could hardly be called 'olfactory'. For practical purposes there is then no 'racial odour'.

SMELL AND SEX DIFERENCES

The question as to whether there is any difference in the perception of smells by men and women has aroused interest for many years and in many countries. At the end of the last century the reports of investigations of this type were coming in by the dozen, all contradictory. E. H. S. Bailey and an American team tested a series of 48 persons with aqueous solutions of oil of lemon, oil of wintergreen, hydrogen cyanide and potassium hydroxide. These solutions were successively diluted until the minimum perceptible was reached. They reported that men were better at smelling and confirmed this in a second series tested with clove oil, amyl nitrite, garlic extract, bromine and potassium cyanide. At the same time, the Italians Ottolenghi and Lombroso had been testing criminals. They found them less sensitive than honest men but the differences between the men and the women were confusing and contradictory. A French team, led by Toulouse and Vaschide, again at the same time, carried out four series of tests. They found that the minimum perceptible odour was 9 : 100,000 for men and 1 : 100,000 for women. In recognition tests for 9 odours: orange flower water, cherry laurel water, rose water, alcoholic solutions of citral, artificial musk, essence of mint, anethole and camphor, women scored higher marks. They also found that among the aged, women were more sensitive, and that little girls were more precocious smellers. This type of report has continued over the past half century. The author after some years of clinical testing found that if abnormalities due to smoking were excluded there was in fact no difference recordable clinically. For practical purposes, and especially as far as medical problems are concerned, there is no difference in the sense of smell between men and women. This is not to say that there are no differences of a finer kind. One point of particular interest is the possibility of partial anosmias and a variable sensitivity to different odoriferous substances not only between male and female, but even during the life cycle of an individual female. This possibility was first raised by the peculiarity of women's attitude to flavours during pregnancy, but no series of tests has been able to demonstrate this as being due to variable sensitivity with any certainty.

The most striking work involving partial anosmias of this type has been carried out in France by Le Magnen and by Guillot from 1948 onwards. The most extraordinary compound tested from this point of view is Exaltolide whose formula is 14-hydroxy-tetradecane carboxylic acid. Most men and most children cannot smell this substance at all, but most women find it very strong and the strength varies with the cycle. Removal of the ovaries may produce anosmia to Exaltolide but treatment with oestrogens will restore the odour.

The reports discussed in the previous paragraphs are embedded in the emotional language of opinion and firmly held beliefs, originating more in feelings than in thoughtful observation. They represent the close association between the olfactory sense and individuals' need to be socially acceptable, to attract the opposite sex and to recognize strangers so as to defend their own. In biological terms this means that smell plays a role in reproduction, in alarm and territorial marking and in maintaining the integrity of the social fabric. Such an important possibility was long suspected but it is only in the last ten years that the manner in which this takes place has been demonstrated. The odorous substances concerned have ben isolated in many animals and in some cases synthesized and have been given the name of ectohormones or *pheromones*. This name was first used to describe sex-attractants in insects but it now has a much broader meaning to include this type of chemical communication.

There is a certain similarity between hormones and pheromones in that both are chemical substances which are formed by glands and which affect the behaviour, development and reproduction of individuals. They are different in other ways, as while hormones are secreted within and by the individual that they will affect, pheromones are secreted by one individual and carry information to others. In this way the physiological state of one individual will directly evoke specific physiological reactions in another. Hormones are not species specific and their use in medical treatment has been largely due to the fact that they can be extracted from the secreting glands of other animals. Pheromones on the other hand appear to be mainly species specific but there is certainly some overlap between related insect species and there is some evidence that this may occur also among mammals.

Pheromones have been found to act in three different ways:

1. By olfaction
The chemical signals given out by one individual are received by olfactory sense organs of the others and behavioural and physiological changes are then set in motion. These appear to be the most common types of pheromone and indeed all known mammalian pheromones are olfactory. The most extensive studies have been carried out on insect sex-attractants for the purpose of manufacturing pesticides.

2. By ingestion
Ingestion pheromones are little known, but this does not mean that they are rare. The most studied is contained in the queen substance of the honey bee. It is manufactured by the queen bee and its ingestion by worker bees prevents the development of ovaries, thus producing a large quantity of immature females. This substance has been synthesized and also used a great deal as a rejuvenator in humans, but without success.

3. By surface absorption
An example is the 'alarm pheromone' in the earthworm, *Lumbricus terrestris*. This worm secretes mucus from segments all along its body, which is said to act as a lubricant as well as a buffer because large amounts are secreted if the worm is placed in a noxious environment such as acid. Ressler *et al.* (1968) have demonstrated that handling, pinching, or an electric shock will stimulate

a copious secretion which will be highly aversive to members of the same species. This means that it has the properties of an alarm pheromone. The surface on which the substance appears does not matter but it is most effective when dry and is not readily soluble in cold water. An interesting finding is that it will remain potent for many months, unlike most insect and fish pheromones. Some rather important implications are raised here because worms have often been used in learning experiments using a T-shaped maze. The worm gets a shock every time it turns into one of the arms and eventually 'learns' to avoid it. It is very likely, however, that the worm avoids it because it is turned away by the pheromone which it has secreted in previous attempts. This situation may exist in other maze experiments using mammals that can produce olfactory pheromones, and much work carried out by psychologists may not now be valid.

There are sex-pheromones secreted by some spiders and cockroaches which have aphrodisiac properties on the male who receives it by contact. A maturation substance is also known to be secreted by adult male locusts which is received by contact by the young and may be responsible for swarming.

Bruce (1970) has classified the type of response which pheromones induce, whatever their mode of action, in the following manner:

1. Releaser pheromones
These produce an immediate reversible response which is operated directly through the central nervous system, for instance by recognition. They can also function through rapidly acting neuro-humoral channels as in the case of the milk-ejection reflex (Cross and Harris, 1952).

2. Primer pheromones
In this type of situation the exteroceptive response involves the anterior pituitary. It is slow to develop, requiring prolonged stimulation and this initiates a chain of physiological effects in the recipient.

3. Imprinting pheromones
The phenomenon of imprinting, though not widely understood, is probably of great importance. Stimulation at a critical period during development (or the absence of appropriate stimulation) may result in a permanent modification of behaviour in the adult.

The study of olfactory response is very difficult. The responses in mammals are not clearly defined although in insects Wilson (1965) has reported nine different types. Until the exact biochemical patterns of these responses are known and can be assayed we have to rely on behavioural methods in mammals. These must involve observation of animals who have had their olfactory sense destroyed so as to provide a measure of comparison with normal animals. As, apart from man, the daily life of most mammals is virtually determined by olfaction, the sudden extirpation of this all-important sense must have extensive repercussion on behaviour even if pheromone activity is disregarded. This naturally creates difficulties for the experimenter, but the evidence which has gradually accumulated over the last few years together with the great advances made in understanding the chemical senses of insects has begun to form itself into a vivid picture of animal behaviour.

INSECTS

Insects depend on many senses to enable them to carry out the functions of living. In their simplest terms these consist of knowing where to go, when to avoid danger, what to eat, which individual to mate with and where to lay its eggs. There is no doubt that many insects can feel the texture of the surface they are standing on and that it may be agreeable or disagreeable to them. They can perceive differences of a chemical nature or 'taste' with their feet as well as their mouths; temperature, humidity and light influence their behaviour considerably. There have been many attempts to demonstrate insect dependence on sound stimuli and it is clear that they can be sensitive to vibration which also carries a sound. The rasping sounds made by crickets and grass-hoppers are certainly perceptible to others and vibrations set up by the wings of a mosquito appear to be meaningful to its mate. It is doubtful, however, whether the perception of these vibratory stimuli could be referred to as 'hearing' and it has indeed been suggested that in some insects harmonic vibrations are set up in the antenae or other parts of the body and that that is how 'auditory' stimuli are perceived.

Despite all these other sources of sensory guidance, smell is probably the most important sense to insects, particularly to those that fly. Even when they use other senses to some extent smell finds an ancillary or often crucial place. An example is the extraordinary system of communication which von Frisch (1950) discovered in bees. These insects can report to the rest of their hive the distance at which they have found an attractive flower bed and the direction in which it lies. This is done by dancing up and down the inside face of the hive, and the direction that this movement takes relative to the vertical corresponds to the direction they are indicating relative to the polarization of the sunlight outside the hive. Even this extraordinarily sophisticated system of communication requires smell, however, if the bees are to choose the flowers from which they have to collect.

This type of insect attraction is generated by the food itself, but often the smells are produced by products which are certainly associated with the food but not directly part of it. Other reflexes using different stimuli may then be required in order to attack the food. The cattle tick, for instance, may remain dormant for months or even many years until a cow may chance to pass beneath it; then the tick will suddenly be re-animated and drop on to its host. The stimulus which sets up this reaction in an insect which may have been quiescent for so long is olfactory. It is the smell of butyric acid which is present in the cow's sweat that is responsible and indeed the tick will attack any object smeared with butyric acid. Once on this object, however, the tick will make no attempt to suck blood unless it has been warmed to the body temperature of the cow. In many instances olfaction may be seen to be an integral part of a whole chain of physiological events brought about by different environmental stimuli.

In choosing a site to lay its eggs, an insect generally depends also on olfaction. This decision-making mechanism can be regarded in two ways:

1. As a specialized form of food attraction, because it is advantageous to the insects' survival to be hatched in an area where there will be food available. It is also likely that it is specifically the female which will find that particular area attractive. The attractiveness of these odorants for the female appears in a large part to be decided by their genetic make-up. Many insects establish themselves on a single species or even variety of host and many attempts to get them to adapt to other types have been only partially successful.

2. As a form of imprinting, because the insect has hatched in a particular environment and its recognition from olfactory or other signals becomes part of its characteristics. This mechanism

probably does occur but it seems to be overshadowed by the genetic structure. It is not easy to decide the role of smell experimentally because in order to observe more clearly defined behavioural changes in insects, they have been reared on media for which they have a distinct dislike. If after some generations this dislike is not altered to a very marked degree, it does not mean that olfactory imprinting has no place in the finer decisions regarding the choice of equally acceptable media.

In the same way as food odours will attract insects of a species in general, and foods which will supply the newly hatched insects will attract the egg-laying female in particular, there is an even more specific attractant. This is the sex-attractant which females produce from their own body and which will attract the male. Here is a true pheromone, a chemical substance secreted by the female which when perceived by olfaction will induce the most characteristic response in the male. So specific in fact that it can be used as a method of bio-assay. A typical sex pheromone and one which is probably the most well known is that produced by the female silkworm moth, *Bombyx mori*, and found in the sacculi laterales of the last abdominal segments. Males react to extracts of this secretion with high excitement and with copulatory attempts on objects impregnated with it. The pheromone itself was isolated and analysed so that its formula is known to be trans-10, cis-12-hexadecadien-1-ol and it has an extraordinarily high biological activity. It is active even in amounts as small as $10^{-12}\mu$g and action potentials can be recorded from the olfactory receptors on the antennae of the males at concentrations of $10^{-3}\mu$g. The males will orient their flight upwind to air-currents bearing this pheromone and will follow the gradient over great distances. This degree of sensitivity can only be olfactory as no other sense can approach it.

The *Bombyx mori* pheromone has even been synthesized and given the name 'Bombycol' and in this form shows remarkable potency.

Other sex pheromones have also been isolated in this way, namely those of the moth, butterfly, fly, beetle and cockroach. That produced by the female gipsy moth, *Porthetria dispar*, has the formula d-10-acetoxyl-cis-7-hexadecen-1-ol and is also active at concentrations of $10^{-12}\mu$g. The synthetic analogues of these pheromones have been produced to be used in pest-control, but clearly such processes of isolation, analysis, bio-assay and synthesis are too uneconomical for use on a large scale. The Mediterranean fruit-fly, *Ceratitis capitata*, was used to screen a random selection of many thousand organic chemicals and in this way sex attractants were discovered much more cheaply by trial and error.

Such extensive work on insect pheromones and further studies which suggest that homing habits in fish may also have an olfactory basis of this type naturally turn our attention towards mammals.

It is the elegant work done by Bruce and by Whitten on experimental animals that has most dramatically demonstrated the importance of pheromones in mammal behaviour. More and more work, however, is now being reported by those concerned with farm animals and animal reproduction so that the place of pheromones in the animal kingdom including primates is now beyond all doubt. The question that remains is their presence and function in human beings. Today this is still speculative, but it would be most unlikely if they were to have no role at all.

MAMMALIAN PHEROMONES

It is olfaction that determines the behaviour and daily life of most animals. This sense guides them to their food, leads them in the direction they must take and, through the mediation of pheromones, has an important effect on their psychological state and hormone production. In order to review these responses we shall, after Bruce, continue to divide them into Releaser, Primer and Imprinting Pheromones; but what is known regarding primates will be discussed in a special section at the end.

1. Releaser pheromones

These produce an immediate, reversible response operated directly through the central nervous system or through rapidly acting neuro-humoral channels. They are responsible for some aspects of sexual behaviour—aggression, territorial marking, homing and possibly recognition of social status.

1. SEXUAL BEHAVIOUR

Again it is to Le Magnen (1952) who had for many years worked on olfactory differences and sex, that we owe the first report suggesting changes in sexual behaviour directly induced by smell. He showed that adult rats discriminate between the sexes by olfaction, and that males can distinguish in the same way between receptive females and those which are sexually quiescent.

It is appropriate at this stage to review briefly the manner in which sexual function is dependent on hormonal factors, as it is probably through the release of these hormones that the pheromones act.

In most species there is a mating or breeding season during which the female will receive the male. The duration of this season varies with the species as does the frequency with which it occurs. In humans there is no such clear-cut season but birth statistics show a period of greater fertility from April to June and this has led to suggestions that a vestigial season inherited from primitive man exists to some extent even at present. Statistical considerations of this type are very slight evidence, as humanity has organized itself into many types of social cycles which run concurrently and which can influence optimum conception dates more than any vestigial cycle could reasonably be expected to do. Holidays are taken at specific times; the school year is broken up into terms, etc.

Apart from this difference between most animals who show seasonal desire and primates who show constant desire, there is another basic one—that the former have an oestrous cycle while the latter have a menstrual cycle.

The periods of sexual activity which occur once or many times in the breeding season of the lower mammals form the oestrous cycle. Some animals, like the bitch, have a single cycle throughout the season and are known as monoestrous. In others like the cat, mouse, rat, mare, cow and sow, the oestrous cycle is repeated many times in the absence of successful coitus. These are referred to as polyoestrous animals.

The oestrous cycle consists of many stages. During preoestrus there is growth and increased vascularity of the uterus with congestion of the vagina, and bleeding occurs. The Graafian follicles of the ovary are undergoing maturation and it is followed by the second stage of oestrus. This is the period of sexual desire in the female and corresponds with the time of ovulation. After oestrus a period of pregnancy or pseudo-pregnancy may occur when the hormones produced by the corpus luteum which has formed in the ovary stimulate the proliferation and activity of the uterine glands together with a great increase in the blood supply of the uterus and growth of the mammary glands. In the time between each breeding season there is a period of anoestrus.

The female primate is not subject to a breeding season and has therefore no oestrous cycle. Instead her physiology is subject to a menstrual cycle. In the human female when the damage caused by the menstrual period to the uterine lining has been fully repaired on the fifth or sixth day of the cycle, a proliferative stage starts. During the following eight days this lining thickens, becomes more vascular and its glands elongate and become dilated in their deeper part. On the 15th day a premenstrual or progestational stage begins with increasing thickening of the lining mucosa (endometrium) which is thrown into folds and a great distension of the glands. On the 28th day the cycle ends and the first day of the new cycle begins with menstruation. The superficial part of the endometrium is shed and considerable bleeding occurs. This menstrual cycle is itself the result of hormonal changes produced by the ovarian cycle.

During the fetal life of the female the ovary is covered by a layer of cells called the germinal epithelium. These cells grow into the substances of the gland, some enlarging and differentiating into ova, which are surrounded by a single layer of smaller cells which become known as the membrana granulosa. The whole is known as a primordial follicle and the ovary at birth may contain between 30,000 and 300,000.

Follicular growth sets in at puberty and near the end of each menstrual period one of these follicles develops. The ovum again enlarges and the granulosa is now formed of many cells which are split into two layers by a growing collection of fluid—the liquor folliculi. At ovulation this follicle, now known as a Graafian follicle, ruptures the surface of the ovary and the ovum floating in the liquor escapes into the perifoneum to enter the Fallopian tube. The collapsed walls of the Graafian follicle then proliferate and differentiate to form large pointed cells containing large nuclei and yellow pigment (lutein). This yellow structure becomes known as the *corpus luteum* and it matures about the 19th day of the menstrual cycle, degenerating just before the onset of the menstrual period. If pregnancy occurs it will last until its termination.

The ovary secretes two hormones, an oestrogen (oestrus–producing) formed mainly by the follicular tissue and progesterone which is produced by the corpus luteum. By means of these two hormones the ovary is responsible for the growth and development of the uterus, Fallopian tubes and vagina at puberty; the development of the secondary sexual characters and sexual psychological differences; changes that take place in the mammary glands and during pregnancy including the embedding of the ovum and development of the placenta.

The oestrous cycle in lower animals and menstrual cycle are controlled by these ovarian hormones. Thus oestrogen alone is secreted during the growth of the Graafian follicle until ovulation. (Fig. 10.1). It gives rise to the proliferative phase. When the corpus luteum begins to produce progesterone as well, the progestational phase is induced. The corpus luteum then degenerates, with the disappearance of both hormones, and menstruation takes place.

There is a close association between the production of these hormones by the ovary and the production of three of the hormones of the anterior pituitary gland. These are:

1. Follicle-stimulating hormone (F.S.H.)

This is responsible for the ripening of the ovarian follicles.

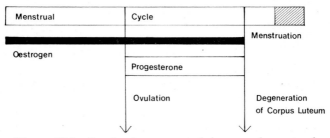

Figure 10.1   Ovarian hormones and the normal menstrual cycle.

2. Luteinizing hormone (L.H.)

This hormone has to be added to F.S.H. for ovulation to take place and the production of the corpus luteum. Its production is stimulated by the rise in the level of oestrogen in the blood which also inhibits F.S.H. production.

3. Luteatrophin, prolactin (L.T.H.)

This initiates the secretion of progesterone by the corpus luteum and maintains it.

The production of F.S.H., L.H. and L.T.H. by the anterior pituitary gland is itself regulated by:

1. The feed-back effect of the blood ovarian hormone level.

2. The central nervous system acting through the hypothalamus. The endings of the hypothalamic nerve-fibres produce chemical substances known as Releasing Factors. These are carried by small

portal vessels to the anterior pituitary. Three Releasing Factors have been described.

1. The luteinizing hormone releasing factor (L.R.F.) This causes release of L.H. after a few hours with the rupture of the ovarian follicle.

2. F.S.H. releasing factor (F.R.F.).

3. Prolactin inhibiting factor.

Ovarian hormone blood-levels affect the production of these Releasing Factors but it is also along that pathway that external stimuli affect sexual function through the nerve-fibres of the hypothalamus.

The male hormonal arrangements are simpler but similar. One hormone, testosterone, is produced by the testis and it is responsible for maintaining all the male sexual characters including emotional make-up as well as his reproductive competence. F.S.H. stimulates spermatogenesis and L.H. is necessary to stimulate the interstitial cells of the testis.

The external stimuli which can affect sexual behaviour through the hypothalamic Releasing Factors are many and include the following:

Auditory stimulation had been shown by Zondek (1967) to induce changes in the oestrous rhythm of rats. An electric alarm bell produced prolonged or persistent oestrus and yet had no effect on immature females or on spermatogenesis in the male. In contrast to a stimulating effect on the sex organs sound had an inhibitory effect on reproductive function.

Visual stimulation also has an effect upon the reproductive mechanism. The female pigeon does not ovulate spontaneously but will do so if she perceives another pigeon or even her own reflection in a mirror. The ferret has been used in numerous studies of the importance of light in the control of sexual periodicity. It is the hours of light to which the animal is exposed daily which stimulate the pituitary into initiating the reproductive cycle.

Olfactory stimuli, it will be shown, play a major role in the control of reproduction among mammals, more especially in connection with the effects on the oestrous cycle and pregnancy.

The manner in which the external stimuli induce the production of Releasing Factors by the pituitary is unlikely to be a simple reflex phenomenon as in many circumstances the effects are long delayed, demanding weeks of stimulation. These complicated and as yet not fully understood mechanisms are summarized in Fig. 10.2.

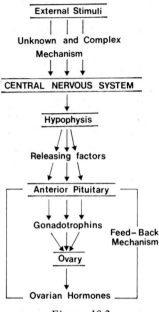

Figure 10.2
Sex hormones regulatory
system.

The place of pheromones in sexual behaviour

The preceding paragraphs on the physiology of sexual function and the manner in which external stimuli can alter it should put the olfactory stimuli into a proper perspective as only a part of a vastly complex system in which alternative pathways can be used to the same end. Furthermore the degree of importance which smell has in influencing sexual behaviour varies from species to species, being the main stimulus in some and a minor one in another. Our requirement is to be able, eventually, to assess its role in human beings where not only visual, auditory and tactile stimuli also play a part but where a large combination of psychological factors derived from many sources is of major importance.

Le Magnen's original work demonstrating the ability of rats to discriminate between the sexes by olfaction was repeated successfully in mice. Michael and Keverne (1968) later showed that the domestic cat will manifest oestrous behaviour if introduced into a cage recently occupied by a male. This does not happen if the cage has been washed.

The effect of anosmia was examined in mice by Whitten (1956) and in guinea-pigs by Donovan and Kopriva (1965). When the olfactory lobes had been removed from the female animals, their cycles continued normally, but despite this they developed considerable disturbances of receptivity. If coitus was performed, however, pregnancy followed its normal course. Franck (1966) showed that if a female rabbit was rendered anosmic before puberty she would show a complete absence of sexual behaviour in adult life.

Because of its economic importance, a lot of the work being done is in farm animals. Signoret and others (1961) have described how the smell of the boar is necessary for the culmination of the immobilization reflex in the oestrous sow. Without that reflex the prolonged act of coition cannot take place successfully. Here smell is the most important requirement but the sight and even the sound of the boar have some contributory value.

The ram discriminates by odour between oestrous ewes and those which are not receptive. With the olfactory lobes destroyed the anosmic ram is unable to recognize the oestrous eye and has to use trial and error. When correct contact is finally established, however, copulatory activity is unimpaired and fertility normal (Lindsay, 1965). The normal ram will find pregnant ewes unattractive but he will be immediately attracted if one is smeared with a swab taken from the vagina of an oestrous ewe.

Young stallions are apparently dependent on odour to recognize their mating partner, but in the more experienced horse the sight of the mare takes precedence.

It seems that the important role of these pheromones is in the initial recognition of the partner and in the integration of behaviour between the sexes. Carr et al. (1965) stated that in rats the attraction of the female rats to males is either dependent on gonadal hormones or on experience, one factor being able to replace the other.

The site of action of the pheromonal stimulus and the sequence of events probably varies in different species; in mammals the male sex pheromones seem to act as an aphrodisiac for the female, while the female pheromones are mainly for identifying the sexual state.

There is no doubt that vaginal secretions contain female pheromone because smearing non-receptive females with vaginal secretions from receptive ones misleads the male; but the source of male pheromones has been in some doubt. Bronson and Caroom (1971) have obtained an attractant pheromone from the preputial glands of the male mouse. They carried out a series of experiments which compared the preference of female mice for various urines and secretions. They preferred male urine to that of females and to the secretions of male glands such as the sub-maxillary, Harderian and lacrymal glands. The urine of males whose preputial glands had been excised was of little attraction, but the saline homogenate of preputial glands was preferred to all urines. The factor was extracted from the gland homogenate with a lipid solvent and remained equally potent.

2. AGGRESSION

The causes of aggression are very important but still mysterious. In humans psychological factors are continuously being investigated and because the implications for political and social as well as individual considerations are so notable, many experiments have been carried out on animals. Some have been behavioural in nature, plcing animals in normal and abnormal situations such as over-crowding, starvation, etc.; others are neurophysiological with sections of the brain, destruction or direct stimulation of the basal ganglia to record the resulting behavioural changes. All this has left a certain confusion regarding the many possible causes of aggression and the physiological manner in which the changes leading to this emotional state are mediated. In adult male mice, however, there is convincing evidence that aggression is released by olfactory signals alone.

An aggression factor is contained in the urine of male mice as aggression can be induced in mice who normally live in harmony by anointing one with the urine of a strange male. Two strange males may be induced to live in peace if their natural odour is masked by a perfume and aggressive behaviour will not set in so long as the scent persists. Excision of the olfactory lobes eliminates all aggressive tendencies from the anosmic animal but it does protect it from normal individuals (Ropartz, 1968). This means that the production of aggressive pheromones is independent of olfactory information derived from the environment.

There appear to be two olfactory pheromones related to aggression in mice :

A urinary factor corresponding to the group odour and which is produced in coagulating glands when mice are placed in close contact.

A plantar factor which is elaborated by the pads of the feet and this characterizes the individual within the group (Ropartz, 1967).

3. STRESS

Stress is closely allied to aggression and the hormonal changes brought about by stress are mediated by Adrenocorticotrophic Hormone (ACTH) and this stimulates the production of cortisone by the adrenal cortex. ACTH is itself produced by the anterior pituitary gland in response to external stimuli transmitted via the hypothalamus. ACTH-Releasing Factor is formed at the nerve-endings and carried by the blood of a portal venous system to the cells of the anterior pituitary in the same manner as F.R.F. and L.R.F.

'Stress', however, is a much less clearly defined entity than any part of the oestrous or menstrual cycle. There are, for instance, similarities and differences between stress caused by physical injury and by psychological or emotional stress such as that brought on by 'alarm'. This makes it very difficult to analyse the manner in which stress situations are represented in physiological terms.

Valenta and Rigby (1968) have nevertheless been able to demonstrate that male albino rats will discriminate between air from the vicinities of stressed rats (S-air) and that from untroubled rats. The stressed rats received electric shocks and the air in which they lived (S-air) was passed by means of a tube to the tested rats. These could be conditioned to demonstrate that they recognized the difference between S-air and normal air by pressing a bar.

4. TERRITORIAL MARKING

An aspect of animal behaviour which is receiving a great deal of attention at the present time is their need to reserve territory for themselves. An obvious manner in which this can be done is by attack, but there must be some system of warning off which might exclude, at any rate, the weaker individuals. In many instances olfaction is directly involved.

The Mongolian gerbil is a good experimental animal in biological research. It is highly exploratory and tractable and requires no water other than that derived as a metabolic by-product. It has a ventral sebaceous gland which secretes a sebum that is oily and has a musk-like smell. Both the male and female 'mark' their territory in a very elaborate manner by rubbing this gland over low-lying objects in the area. They approach an object rapidly, sniff it, and then mount it, pressing their gland on to its surface and then dismount forwards. There are some clear-cut differences between the male and the female. The male gerbils have much larger glands and they mark objects twice as frequently as the female; but if a male enters a previously marked territory, its own marking frequency, which is generally always repeated, is reduced considerably.

Thiessen and his colleagues (1968) have shown that the integrity of this gland in the male is dependent on the gonads and that if males were castrated they gradually stopped marking. This ability could be restored by the administration of 80 μg of testosterone proprionate, but if 640 μg was given, a much more rapid effect was obtained.

A few important facts emerged from this experiment. Both the marking frequency and the size of the gland and its integrity are dependent on the male hormone, testosterone. On the other hand, there is a relationship between the marking tendency of the animal before castration and its behaviour after testosterone administration but this, in turn, is not related to the size of the gland before or after. That means the decline and restoration of markings are more dependent on gonadal secretion than on the integrity of the gland.

It appears, therefore, that although both the gland and the marking frequency are dependent on systemic androgen titres, it is the loss or gain of hormone at some central nervous locus which is the factor responsible for the animal's territorial behaviour and is independent of the gland *per se*. Indeed, some of the gerbils that had received 640 μg of testosterone proprionate had become 'super-markers' but the terminal gland size was nothing out of the ordinary.

This relationship between male hormone production by the testis and acquisition of territory is an interesting one. The comparable association between the male gonad and sexual behaviour as well as aggression has been known since earliest times and castration both in men and animals has been so widely used that there is no lack of descriptions confirming its effect.

Androgen production by the testis, therefore, appears to control reproductive competence, intra-species competition, as well as territorial formation and signalling. As these provide the three main dimensions of species survival it appears that a high titre of androgen would qualify an individual to pass on its genes more effectively than a low one.

5. HOMING
Some evidence is now available that homing by fish and other animals may be due in a large part to a Releaser type of olfactory pheromone.

6. RECOGNITION OF STATUS
Again, the problem of status within a group has fascinated students of human and animal behaviour principally because of interest in the status-structures which exist in human societies. It is probable that in some animals recognition of status is based on olfactory information.

2. Primer pheromones

The response which these pheromones induce involves the anterior pituitary gland. It is slow to develop, necessitating prolonged stimulation and initiates a chain of psysiological effects in the recipient.

Primer pheromones are implicated in a number of bodily functions in various animals.

1. CONTROL OF OESTRUS
In most mammalian species the male appears to exert a controlling influence on oestrus and a major part of this influence is exerted through olfactory stimuli.

It is well known to farmers that among seasonal breeders like sheep and goats the introduction of the male shortly before the start of the breeding season stimulates ovulation and heat in the female and so terminates anoestrus. In this case the influence of the male reinforces other environmental stimuli and is clearly not the only one concerned. In order to be effective the male must present a novel stimulus as, if he has been continuously with the females or if he is introduced after they have become cyclic, no further effect can be obtained.

In the mouse the oestrous cycle is susceptible to environmental conditions and is greatly modified by olfactory stimuli from both male and female.

When a female is isolated, her cycle is 5 to 6 days long but irregular. If a number of females are brought together in conditions of communal living there is a mutual suppression of oestrus. In small groups of females pseudo-pregnancies are common and in large groups anoestrus is likely to occur. The introduction of a male has an immediate effect in stimulating a new cycle so that oestrus becomes synchronized in the group. In the isolated mouse the oestrous cycle is shortened and coitus is concentrated on the third night. These important effects were described by W. K. Whitten.

Olfactory stimuli are responsible but the influence of the male is transitory and is exerted on a single cycle only unless it is renewed every cycle.

This 'Whitten effect' occurs also in sheep and goats but only at the start of the breeding season. It varies also in other species; rats have a more stable cycle, so smaller numbers are affected, and hamsters, for instance, are not affected at all.

2. EFFECTS ON PREGNANCY

The extraordinary degree of olfactory control which male mice exert on the female has been demonstrated by Bruce (1959, 1960). The mated female will remain vulnerable for as long as four days after coitus. If she is put in the presence of a strange male the pregnancy will fail and she will return to oestrus. Housing the female in a box previously soiled by a strange male will have the same effect of interrupting the pregnancy but if the pregnant female has been made anosmic she will be unaffected even by close approximation to a stranger. In other words olfactory impulses can block a pregnancy with all that this involves. This blocking effect, known as the 'Bruce effect', has not been demonstrated in many animals and in particular in farm animals, but it has shown in a most dramatic manner how powerful olfactory stimuli can be.

3. ENDOCRINE CHANGES

The Bruce effect, or pregnancy block, reflects a failure of implantation and this in turn is the result of a failure of the luteotrophic activity which could normally be expected after coitus. The administration of prolactin or its endogenous production by ectopic pituitary grafts or by a suckling litter all give protection from such a block. This is likely to be by interference with hypothalmic functions and it is interesting to note that similar protection can be obtained by the administration of reserpine which is known to depress hypothalamic activity.

The male mouse pheromone acts therefore by stimulating the hypothalamus. This in turn by the humoral releasing factors induces the anterior pituitary to produce F.S.H. and lutenizing hormones so that oestrus and ovulation is brought about and at the same time prolactin production is inhibited and thus implantation of the ovum is unprotected.

The female pheromone which seems to be produced by grouped females has the opposite effect.

Avery (1969) has studied the pheromone-induced changes in the concentration of anterior pituitary acidophil cells and thus in L.T.H. secretion. This was done in the glands of grouped females, as well as those exposed to restrained males for the first and second nights and to released males on the third night, following removal from colony housing. Acidophil granules in anterior pituitary cells represent L.T.H. so that a very high acidophil concentration in the cells can be considered to indicate the suppression of L.T.H. secretion by these cells, while a declining concentration which suggests an increasing discharge of granules represents increased L.T.H. secretion.

In grouped females, that is those exposed to female pheromone, there is a maximal acidophil concentration which means that their anoestrus is not caused by continuous L.T.H. secretion. This concentration is greatly reduced after exposure to the male so that the investigation supports the concept that the anoestrus caused by female pheromone is due to a failure of F.S.H. secretion. Subsequent exposure to the male pheromone stimulates F.S.H. production and this in turn leads to an increased production of ovarian oestrogen. The rise in oestrogen blood levels then causes a suppression of F.S.H. production and, reciprocally, promotes L.T.H., thus setting off the cyclical behaviour of the normal female.

4. SOURCE AND CONTROL OF PRIMER PHEROMONES

This problem is still entirely speculative. The questions that must still be asked in a general sense are many. Which glands or organs secrete these sex-pheromones? Are these odorous substances of multiple use? Can a pheromone which attracts a female of the same species produce an aggressive response in another male or warn a younger one away from a marked territory? Can a pheromone act as both releaser and primer and then imprint on the newly born? Is there a different pheromone for each function? At present we are far from giving constructive answers to this type of question. The fact that pheromones, unlike hormones, are species specific as far as we can tell makes it, of course, more difficult to study their effects; but it does appear that they are controlled by the gonads. Castration will make the malodorous character of the billy-goat disappear and the various 'effects' described can be lost after castration, but acquired by ovariectomized females treated with androgen.

The glands which produce known odorous substances are diverse.

1. The preputial glands. Those of the musk-deer, musk-rat and beaver produce odours well-known to man and used extensively in perfumery. There can be little doubt that they are important to their own species where they are likely to represent pheromonal activity. The preputial secretions of the mouse and rat which have been shown to contain pheromones have little odour to the human nose, but this is not to say that they are not strongly odorous to their own species.

2. Anal and perineal glands produce odorous secretions in the Mustelidae and are especially well-known in the skunk, the civet, the stoat and the ferret. They produce secretions in many other animals including the rabbit, and those of the male guinea-pig at least have been known to regress after castration.

3. Sebaceous gland secretions are odorous in the hamster, shrew, mouse and billy-goat. The ventral gland of the Mongolian gerbil is, of course, of great importance.

4. Apocrine glands can exude a strong odour and are particularly important in the forearm of the monkey and the axilla of man.

5. Faeces. It is not impossible that some of the odorous components of faeces have pheromonal quality.

6. Urine. Many people claim to be able to tell the difference between male and female urine and in 1950 the isolation of the steroid Δ^{16}—androstenol from human urine (Brooksband and Haslewood) generated a lot of interest because of its musk-like odour. Whether it has any pheromonal activity is very doubtful.

Some elements of urinary smell are due to secretions of nearby glands like preputial glands, but as it has a smell of its own due to the metabolic products it contains, urine taken directly from the bladders of male mice has been used to see if they had sex pheromone activity. This turned out to be the case.

7. Seminal and vaginal secretions have a marked and specific odour and are likely to be of considerable importance.

3. Imprinting pheromones

The subject of 'imprinting' has attracted a great deal of attention but its discussion lies outside the scope of this book. It is a term used to describe a disturbance of adult behaviour which results from environmental inadequacy during development. This inadequacy has to take place at critical periods and may result either from the appearance of an abnormal stimulus or from the absence of an appropriate stimulation. The disturbance of behaviour to which it may lead may be permanent.

Visual stimuli have been shown to affect many animals; particularly dramatic has been the demonstration of attachment to another species of birds hatched by a stranger—even a man. This type of alteration of behaviour has been caused by olfactory stimuli in fish as far as homing is concerned and there is now some evidence of this occurring in mammals.

The social behaviour of mice and rats can be modified by olfactory experience during suckling (Marr and Gardner, 1965). Female mice reared in the absence of the male adult show a loss of sexual discrimination when they grow into adult life. This type of disturbance may also be induced if the atmosphere has been heavily scented so that the natural odours are masked.

THE ROLE OF PHEROMONES IN PRIMATES

Because of their close relationship to man, the presence of pheromones among primates would be of considerable importance and many efforts have been made to demonstrate their existence.

Many prosimians have apocrine glands specialized into producing scent which these animals appear to use for territorial marking, for self-marking and for marking of the female by the male. Ring-tailed lemurs, for instance, depend on a number of olfactory signals to maintain their social integration, at least when observed in captivity (Evans and Goy, 1968).

Marmosets have specialized perineal and genital glands that become functional at puberty and produce a smelly secretion which appears to have some importance in sexual behaviour, while the toque macaques have a vaginal secretion that is so strong in smell that it can easily be perceived by humans. Routine scenting of the female ano-genital region by the male is the normal behaviour among macaques.

In most primates there is a rhythmic variation in the sexual interest of the male corresponding to the female's menstrual cycle. This has been described by Michael (1965) in the Rhesus monkey which is a microsmatic higher primate. These variations are abolished by ovariectomy but the sexual behaviour of the male can be restored by injecting ovarian hormones into the female. These findings provide clear evidence that the males are aware of their partners' endocrine status. When progesterone is injected the sexual activity of the pair declines. Two distinct mechanisms appear to be responsible for this decline: there is a decrease in the receptivity of the female indicated by increased refusals of male mounting attempts; the second depends on a loss of attractiveness of the female as indicated by the relative failure of her sexual invitations to stimulate male mounting behaviour (Michael *et al.*, 1967). While the first mechanism seems to depend on a change in a hormone dependent mechanism within the brain of the female, the second requires the mediation of distance receptors in the male. Another fact which has arisen from Michael's important work is that if oestrogen is administered intra-vaginally to the ovariectomized female, it has a greater effect on the male partner's sexual behaviour than if it is given sub-cutaneously. The intra-vaginal dose may be too small, however, to affect the female's receptivity. We may say, therefore, that there is an oestrogen dependent vaginal mechanism which is capable of influencing the male independently of the female's receptivity.

In other series of experiments Michael and Keverne (1970) again used rhesus monkeys where the male had been trained to press a lever which would open a trap door if he wanted the female placed in an adjoining cage. As expected, they were not interested in ovariectomized females unless oestradiol was placed in the females' vagina. The effect of the oestrogen, however, was cancelled if the male was made anosmic. Michael and Keverne then tested the effect of transferring vaginal secretions from normal females to the genital skin of overiectomized females. Their male partners, previously uninterested, immediately began to press the lever which gave them access to the females.

It appears from these experiments that there is a component in the vaginal secretions of the normal female rhesus monkey which provides an olfactory stimulus which induces sexual activity in the male. Although not yet isolated, Michael and Keverne have suggested 'Copulin' for its name.

There is no doubt that Michael's work has important implications, as it has been carried out in a microsmatic primate.

HUMAN SOCIAL IMPLICATIONS

It is customary to state that 'apart from man' the animal kingdom is dependent on the sense of smell. Most of the accumulated knowledge which has led to this type of generalization is the understanding of the place of smell as an aid to *recognition*. Animals recognize the food they are to eat mainly by olfaction and in the same way they recognize the place where they must lay their eggs. They recognize members of the same species and the more subtle differences which mark members of the opposite sex. In this function of acting as a simple 'signal' it is self-evident that olfaction has a very minor role in man.

The discovery of the pheromones and the demonstration of their power in mammals and particularly in primates poses questions of an entirely different nature. Here we have a chemical secreted unwittingly by one animal exerting a direct influence on many aspects of the emotions and behaviour of others. The important thing is that the recipients are unprotected by consciousness even though they may be aware of subtle odour changes. It is as though there is a whole subconscious network of stimuli and responses controlling some of the emotions and feelings of animals, and when we consider that these are concerned with such things as aggression, territorial marking and reproduction, the place of pheromones in animal psychology becomes, if anything, dramatic. The implications that this has for man is that we may be as unaware of our control by pheromones as we are of the effects of our own hormones. It should be made clear, however, that no valid evidence proving the existence of human pheromones is available at present.

There is a great deal of circumstantial evidence on both sides. In the first place the important signals which men use to communicate with each other are verbal, so that there is not only an exchange of information, but of abstract thought, while poetry and the unique example of music allow the communication of feeling in a largely non-verbal manner. Visual contact is very similar and in the situation of writing there is combination of visual communication and verbal which is an extension of auditory. This is all brought about by a conscious, voluntary effort and in this, other senses such as touch play a secondary role and it is evident that smell plays a very minor one indeed.

There is another system of communication among men which is at a somewhat less conspicuous level, and although it can to some extent be simulated by voluntary effort, to a large extent it is involuntary and even unconscious. This is the sort of information usually associated with feeling such as anger, pleasure, mockery etc. that is imparted by the tone of voice, the facial expression, changes in skin colouring and so on. As some aspects of this system are entirely conscious and used freely almost at will while others are entirely unconscious, it is likely that there is a whole range of signals of a varying degree of consciousness, while more sensitive people may be aware of a greater range than less sensitive ones. If smell, and particularly pheromones, have a place it must be at a less conscious level of this system.

Of course, following the 'apart from man' phrase that is always repeated in any work on olfaction, the questions have to be asked: is there any function for smell at all? Are we underestimating the value of smell to man? We know that man can perform all the important functions in life in the absence of olfaction, but this does not necessarily mean that the sense of smell is not given an important role when it is present.

Looking around us, we see a civilization which is not only perfused by scents of every kind but where they are created and produced by man, and often used in a manner not far from obsessional.

We live in a world where incense has had a crucial place in religion and ceremonial and where evidence of the use of perfumes stretches back to pre-historical times. This aspect of human asociation with odour is discussed in Chapter 11, but in the world of today we see a vastly and rapidly increasing use of perfume. Scent is built into goods of every description and particularly in plastics where this can be done more easily. Soaps, detergents and cleaning materials all have added smells, but more important is the personal use of perfume in the form of toilet preparations. Both men and women use creams, lotions, shaving soaps, powders and unguents of different kinds. The increasing use of 'deodorants' is perhaps the most ominous, as they not only reduce body odour but they mask it with scent. Perhaps they should be referred to as 'odorants'. Most of the perfumes contain musk, civet or their synthetic analogues—known animal pheromones—to give them 'body'.

In such an atmosphere it would be ridiculous to suggest that olfaction has no place in human relations. We have only to read the advertisements on the boardings to see the claims made for its role in social integration. This does not prove, of course, that human pheromones exist, but the circumstantial evidence is two-fold: first the olfaction is an important sense in man, used extensively socially as a means of communication between individuals; secondly pheromones have been shown to exist in most species including microsmatic primates.

If the possibility that pheromones may exist in man is accepted, the social consequences should be considered. The psychology of sex-attraction and of sympathy and antipathy should include olfactory stimuli, and the possibility of imprinting pheromones being responsible for differences in behaviour—between breast and bottle-fed infants, for instance. In therapeutics the effects of drugs on smell should be studied, particularly hormonal compounds like the contraceptive pill.

The most interesting question should be the social effects of our increasing odourization. Particularly now that even vaginal deodorants have been put on the market, is there a danger to the social fabric of masking an important vaginal pheromone? There are three possible ways in which we may be interfering with our olfactory environment:

1. Enhancement

Toilet preparations such as make-up, if properly used, tend to enhance the attributes and thus the attractiveness of the person. Even changing fashion provides for the important human need for novelty. In the same way perfumes, which usually contain musk-like chemicals, render the individual more attractive and have been selected over centuries. If this is the case, there appears to be only biological advantage to be gained from their use.

2. Suppression

It may be that scents counteract and mask all the natural pheromones. This can be considered as an unmitigated disaster or, on the contrary, as the only way in which the inter-personal relationships of a civilized society can be maintained. Scents could damp down disruptive sexual urges, suppress aggression and diminish territorial impulses in people who have to move about in crowds and who often live on top of one another.

3. A mixture of both

This is probably the case, with some enhancement obtained through trial and error and some necessary or unnecessary suppression. The possible danger is that it occurs in such a haphazard manner and deodorization may carry its own momentum regardless of the effects, as in the case of vaginal deodorants.

CHAPTER ELEVEN

Perfumes, Flavours and the Control of Smell

PERFUMES AND SOCIETY

From the earliest recorded times, man has been familiar with a multitude of perfumes. The word is derived from *per fumum* which implies an odour produced through smoke, associating it closely with incense.

Theophrastus, one of the earliest writers on perfumes, described in great detail the various aromatic parts of plants and the oils used to digest them, explaining that these facts had been known as far back as man remembered.

Social uses of perfume are well demonstrated by the Ancient Egyptians. First they were used as offerings to the Gods and formed a part of the religious ceremonial at least as important as that provided by the auditory and visual stimuli of holy chants and symbols. It is likely that the emotional content of the odorous part of the service was of prime importance and often formed its centre piece. This is dramatically shown in a large tablet on the breast of the Sphinx where King Tothmes IV (about 1600 B.C.) is seen offering incense and fragrant oil or unguent. The second use was aesthetic, as they used perfumes extensively together with other cosmetics in order to make themselves more attractive. The third use was specific to the Egyptians and has not been important elsewhere: that is in embalming the dead.

This double function of odorants, the sacred and the profane, has existed in all cultures and one has not impeded the other. It was only with the advent of Puritanism that a direct attack on the cosmetic use of perfumes was made and it is interesting that incense burning was also excluded from religious ceremonies. Since then there has been an uneasy balance in England between the 'sinful' and the beneficial aspects of perfumery. In 1770, according to Septimus Pierse, a Bill was placed before the English parliament which contained the following section:

'That all women of whatever age, rank, profession or degree, whether virgins, maids, or widows, that shall, from and after such

Act, impose upon, seduce and betray into matrimony, any of his majesty's subjects by scents, paints, cosmetic washes, artificial teeth, false hair, Spanish wool, iron stays, hoops, high-heeled shoes, bolstered hips, shall incur the penalty of the law in force against witchcraft and like misdemeanours and that the marriage, upon conviction, shall stand null and void.'

There have been various attempts throughout history to ban perfumes; from Solon of Athens to the French Revolution and to Hitler's Germany. In Britain, at the beginning of World War II, an attempt was made to limit the use and production of cosmetics as 'non-essentials', but this policy was quickly reversed and the role of the industry in 'maintaining morale' was felt to justify the widespread distribution of perfumed substances.

It is very difficult to discover when man first started using fragrant unguents on himself or provided burnt offerings to his God. Paleolithic man painted himself black and ochre. It demonstrates a knowledge and interest in colours which suggests that he may equally well have painted himself with vegetable perfumes, although no containers for them have been discovered.

The use of odours as offerings probably originated in animal sacrifices where the smell of burning flesh was quite naturally held to be 'satisfying' to the Gods. Gradually scented woods were used and then aromatic gums, resins and oleoresins were added to sacrificial fires so that in later times the smell of burning flesh from the sacrificed oxen came to be thought obnoxious to the Gods and had to be disguised by incense. The use of incense became highly ritualized. In Heliopolis, the centre of sun-worship, it was highly complicated; resins were burnt at dawn, myrrh at noon and a special mixture called kyphi at sunset.

Throughout the Middle East the religious use of incense increased. Among the Assyrians it reached the same excesses as so many of their actions—at the feast of Baal over 1000 talents in weight of frankincense were burnt in Babylon. The Jews, however, were late users of incense and the point was never reached in Levitical ritual when incense actually replaced the burnt sacrifice.

In both Rome and Persia burning of incense was an important part of public as well as religious ceremonial. Zoroaster recommended the burning of perfumes on a sacred fire five times a day, while devout Roman families lit daily incense-burners to their *lars familiaris*.

It has been difficult to discover when incense was first used in the Christian church. Theoretically it should be from the very beginning as it was used by its two predecessors the Jews and the Romans, but it is not mentioned in the Apostolic Constitutions and there is no comment by either Cyril of Jerusalem or Augustine. It appears later, however, in the Apostolic Canons (Canon 3) where it is considered one of the requisites of the Eucharistic service. The smoke is now held, in the Catholic church, to symbolize the prayers of the faithful and the virtues of the saints.

The Church of England has always been ambivalent towards incense, banning its use at times, permitting it at others. Although the official attitude is usually against its use during services, it has continued to appear despite the contrary opinions of the Archbishops of Canterbury and York.

Although at the present time perfume, in the form of the cosmetic industry, is so distant from incense as to make us wonder where the association lies, in early times the same substances were often used in both products. The simplest perfumes were made by digesting odorous vegetable substances with sesame, almond or olive oil—a process little different from the 'maceration' used at the present time. Toilet articles and containers for perfume have been found from the first Egyptian dynasty, dating back perhaps to 5000 B.C. Jars and vases containing unguents have frequently been found in tombs, among the most beautiful those in the tomb of Tutankhamon discovered by Carter in 1922. One vase contained an aromatic ointment dating probably from 1350 B.C. The perfume seemed to some people like broom, to others like valerian, and yet others compared it to coconut oil. Chapman and Plenderleith analysed it in 1926 and found that it consisted of a neutral animal fat base with a resin or balsam. This was probably Indian nard—the spikenard of the Bible—which is in fact allied to valerian.

Another find rich in odours was the tomb of Queen Hetepheres, mother of Cheops. The scents which emanated from the numerous containers were similar to those popular in the East today.

In Ancient Egypt, perfumes had an extremely important place. No ceremony, religious or public, no feast was without the extensive use of perfumes. Aromatic oils were sprinkled on visitors' heads, aromatic fumes were produced in numerous burners. Fresh flowers were everywhere, as well as wreaths, and lotus and jasmin chaplets were placed round the necks of guests. This is reminiscent of the heavy necklaces of 'foll' sold in every open cafe and even through the windows of cars in Cairo and Alexandria today. In the dry coolness of the evening the air feels heavy with their scent.

The odorous constituents of the perfumes were usually thyme, origanum, myrrh, olibanum and spikenard. They were mostly macerated in sesame oil but also in almond and olive oils. But the use and variety of perfumes reached its climax at the time of Cleopatra and it is certainly worth reading Plutarch's version of the seduction of Mark Antony!

The use the Jews made of the baths and perfumes they brought out of Egypt was predictably hygienic and highly mystical. Right from the start (Exodus XXX, 23 to 25) Moses was commanded to make a fragrant oil for holy anointing. The spices and proportions were precise: 500 shekels of pure myrrh; 250 shekels of sweet cinnamon; 250 shekels of aromatic seed; 500 shekels of Cassia macerated in one hin of olive oil.

The ceremony of holy anointing throughout subsequent history derives from this. It was forbidden to use oil perfumed in these proportions other than for sacred purposes. An extremely sacred perfume was also carefully ordered and prepared (Exodus XXX, 34 to 38) consisting of the following substances:

Stacte—This is probably the oldest aromatic gum we have. According to Pliny it is the spontaneous exudation of the myrrh tree.

Onycha—This is the operculum of a sea-snail which is still used by Nubian women.

Galbanum—A gum-resin which was also supposed to have medicinal properties.

Frankincense—This also was a gum-resin compound. It was imported from the land of Punt and according to Pliny there were 3000 families who knew the sacred trees and guarded them. Because of its rarity and importance it is quite possible that there was some sort of monopoly in its production. Frankincense was also believed to have medicinal properties such as acting against leprosy, tumours and ulcers and as an antidote to hemlock.

These substances were mixed by Moses in equal parts, pulverized and kept in the House of the Lord. It is probably this sacred perfume which was brought out by the contenders in the ill-fated revolt of Korah, Dathan and Abiram against Moses (Numbers XVI, 32 to 35). Two hundred and fifty men were destroyed by providential wrath after that episode.

Women used perfumes as a purifying ritual. Oil of myrrh during the first 6 months and other oils and unguents for the rest of the year. This type of preparation is well described in the Book of Esther (II, 12). It was only after twelve months of such purification at the palace's expense that Esther and the other young girls were allowed to present themselves before Ahasuerus (the classical Xerxes), who wished to replace his repudiated queen, Vashti. This ritual purification was of course also carried out in a similar manner by women of all the nationalities forming the Persian Empire at that time. The perfumes listed in the canticles are the following:

Camphire—from the henna shrub;
Saffron—from the dried stigmata of the crocus sativus;
Calamus—the rhizome of the sweet flag;
Cinnamon;
Cassia—from the bark of a species of laurel;
Aloes—a wood attacked by a disease ('fragrance sinking under water').

The holy and carefully regulated perfumery of the Jews is modest and provincial compared with the grandness of the followers of Baal and Astarte. The court of Assurbanipal was reputed for its great feminization and for the suffocating atmosphere of its aromatics and the smoke from pyres of fragrant woods. Babylon and Niniveh became the centres of the world perfume trade and the position of that part of the world regarding odorous substances persisted till very recently. The Greeks who took control of these middle eastern empires surpassed their predecessors in use of perfumes and what had now become open feminization. Guests of Antiochus Epiphanes, King of Syria, were sprinkled with rose-water by 200 girls brandishing golden watering cans, while boys in purple tunics carried gold dishes containing frankincense, myrrh and saffron.

During the Games at Daphne, each guest entering the gymnasium was anointed with perfume from 15 gold dishes, each containing a different aromatic such as saffron, myrrh, frankincense, spikenard, cinnamon, lilies, fenugreek and amaracus.

Many of these perfumes were imported and often their origin was mysterious and associated in people's minds with wonderful legends, like that of frankincense and its 3000 families. The strangest stories referred usually to the land Sheba, a fruitful source of perfumes. Cinnamon was said to be made from the ground-up nests of the phoenix, that mythical bird said to arise from its own ashes; the cassia plant was believed to be defended by giant bats which made a speciality of tearing out the eyes; while calamus apparently came from the birth place of Bacchus which was protected by winged serpents. It is quite likely that in an age not covered by patent rights these stories were deliberately circulated with the intention of discouraging competition and transplantation of the original plants.

Rome became mistress of these regions after the conquest of Magna Graecia and the subjugation of Antiochus. Although they had become familiar with the use of incense after they had overrun the Greek-populated south of Italy, they were revolted by the feminized, perfumed society which they found. In 188 B.C. the stern Roman Republic of that time published an edict forbidding such usage.

It is not surprising to note, however, that by the time of Nero (A.D. 54) the attitude had markedly changed. According to Suetonius (Book VI) more incense was consumed at the funeral of Poppaea than Arabia could produce in 10 years. Although Suetonius cannot be entirely believed and, indeed, Pliny while telling the same story (Book XII, Chapter 41) refers only to one year's output, it is clear that things had already changed from the days when Rome was composed of aescetic, rough-cut farmers. As the Empire became more Asiatic in nature, so the use of perfumes increased. The excesses of Heliogabalus and his feminine habits (Gibbon's Decline and Fall of the Roman Empire, Chapter 6) more than compare with anything seen in the temples of Astarte or the court of Antiochus.

The perfumes used by the Romans were prepared in three forms:

1. Solid unguents—ledysmata. These contained usually a single perfume such as almond or rose;
2. Liquid unguents—stymmata. The odours were often mixtures of flowers, spices and resins;
3. Powder perfumes—diapasmata. As so many of the products required had to be imported from outside the confines of the Empire, there is no doubt that they must have contributed to the chronic balance of payments deficit, which some historians claim to have been the root of Rome's economic problems.

When the Arabs were able to replace the Eastern Roman Empire as the dominant force south of the Mediterranean, perfumes were cherished in an almost affectionate manner. It was a society which had tried to eliminate sensual excesses, not by punishment and moral guilt, but by legalizing the main sources of licence and desire. In such a society scent is not considered an excitant so much as a comfortable stimulator of the senses. This vast difference in attitude towards perfumes is worth considering in its wider social context as our own society is poised between the traditional Judeo-Christian suppression of socially disruptive desires and a new outlook. There is a certain similarity between the world of Islam and the belief that permissiveness towards sex, for instance, can free desire from its socially dangerous powers. Thus, in Islam the dangers of coveting a woman, the social implications of adultery and the powerful psychological forces engendered through sexual frustration were all by-passed by polygamy and morally acceptable concubinage. This was sexual permissiveness carried almost to its paralysing negation, if only so far as men were concerned.

This same attitude applies to all the senses. Contemplation of beauty in every form to its sensual limit received such complete social encouragement that it would be foolish to use the word excess. Music developed very slowly; listeners would sit still for hours absorbing a repetitive system of melodic patterns and rhythmic modes to be tickled almost into ecstasy by the trills, graces, slides and shakes which occasionally adorned them. The sense of smell took precedence before all the others. Mohamed himself confessed to three favourites—women, children and perfume—in that order. He is even reported to have claimed a miracle: 'When I was taken up to heaven some of my sweat fell on the earth and from it sprang the rose, and whoever would smell my scent, let him smell the rose.'

Rose-water, then, remained the favourite so that when Salah ed-Din (the Saladin of the Crusades) entered Jerusalem in 1187, he had the walls and floor of the mosque of Omar washed with it. The Persians were the greatest producers of rose perfume and their love for its scent is so well expressed in Sâcli's beautiful poem, the 'Gulistan' or Garden of Roses.

When the Arabs had learnt chemistry from the Greeks they began to make great contributions to the art of perfumery and it has been shown that they were familiar with distillation. The quality and variety of perfumes increased considerably between the 9th and 12th centuries as a result.

The *Thousand and One Nights* contain many references to perfumes and some rather strange concoctions are described in Shiekh

Nafzawi's *Perfumed Garden*, intended to free women from un-
wanted body-odour (Chapter XVIII).

Red myrrh pounded and kneaded into a paste with myrtle water.

Lavender pounded and kneaded with musk-rose-water.

Carobs (the fruit of the so-called locust tree) freed from their
kernels boiled with the bark of pommegranates.

Antimony and mastic. The word which is usually translated as
Antimony here is the arabic 'hadida' which sometimes means certain copper compounds which were used to dye hair black.

All these substances were recommended for rubbing over the
body and said to be very dependable.

The use of perfumes even extended to cookery and Abdel Latif
describes a most extraordinary pie. The recipe contains oil of
sesame, ginger, cinnamon, mastic, nutmeg, and rose-water infused
with musk.

Naturally in that atmosphere there was nothing specifically
feminine about perfumes and scents were equally prized by both
sexes.

It should be added that when these perfumes had begun to spread
to Europe, men used them quite uninhibitedly, particularly in
Elizabethan times. Under the Commonwealth they were frowned
upon together with many other adornments but they quickly resumed a continually rising popularity. The production of perfumes
in Europe is closely associated with the growth of the French town
of Grasse. Surrounded by some of the most extensive flower gar-
dens in the world and developed by the growing technology of the
last 150 years, perfumes have reached an incomparable degree of
sophistication, standardization and subtlety.

Not only do people, both men and women, use an increasing
number of highly scented cosmetics such as soaps and shaving-
creams and eau-de-cologne, but perfume is added to every box
of detergent powder and to every furniture-polishing liquid. Perfumes are now being introduced into plastic materials so as to give
them the smell of leather or of wood, as required. There is no doubt
that we are entering a highly aromatic world and when we learn
that even fifteen years ago, in a single month of 1958, the British
sales of perfume alone, excluding cosmetics were valued at
£4,493,000 it is worth remembering that one of the earliest references to trade we have relates to the perfume trade in the story of
Jacob and his son Joseph.

THE MANUFACTURE OF PERFUMES

Perfumes fall into three groups:

(1) Those originating from animals.
(2) Natural vegetable perfumes.
(3) Synthetics.

The past has shown us how individual perfumes and then simple mixtures appeared. Today hardly a perfume is without one substance at least from each of the three groups. Even a simple flower scent will contain an animal or synthetic odorant to give it 'body' and to 'fix' it.

The perfumes from the animal kingdom

These are, in fact, the most vital ingredients of most good preparations. The Greeks gave them a divine origin: Venus invented them and the secret was given to Paris by a nymph and thus found its way to Helen of Troy. They are four in number: Ambergris, musk, civet and castor.

1. AMBERGRIS

The name is due to a mistaken belief that this was a form of grey amber. It has been popular for a long time but has always been very expensive and ten centuries ago it was classed with black slaves and gold as the most lucrative trading products.

It is, in fact, not a resin at all, but a calculus which is produced in the intestines of the sperm whale. The calculus probably results from irritation of the intestinal wall by the indigestible, horny beaks of cuttle-fish which have been ingested by the whale.

The pieces are eventually evacuated and are found washed up on various coasts. Now that whales are hunted pieces are occasionally found in their intestines and are considered a very valuable bonus. The largest piece recorded which has been found in this manner weighed three hundredweight.

The smell of raw ambergris is somewhat earthy with seaweed qualities. In order to use it in perfumery it has to be extracted in alcohol and then allowed to 'mature' for some time. This gives to it a certain velvetiness and endows it with the lasting quality typical of animal perfumes. It is still believed by many to have aphrodisiac properties.

2. MUSK

The active principle of this secretion is a ketone which is itself a biological conversion of oleic acid. It accumulates in the preputial follicles of the musk-deer of the Himalayas and Atlas mountains so that about 2 oz can be collected from a young buck. As these animals are shy and live at a height of 8000 ft there are considerable difficulties in catching them, many being killed in the process, and this has tended to deplete the species. The follicle is referred to as the pod and is a spherical sac about 2 cm in diameter, hairy on one side and with a small orifice.

The smell is slightly ammoniacal and its most important property is the remarkable diffusive powers which it has. It is said to give 'life' to a perfume.

Musk-rat musk has only a limited use.

3. CIVET

The active principle is again a ketone similar to musk. The ketone is also derived from oleic acid.

It is a glandular secretion contained in two small sacs in both the male and female civet cats, although that from the male is better. The cats have a grey fur with black spots and are found chiefly in Abyssinia, although they are also bred in Senegal and Guinea. There is a different species which also produces civet in Indonesia and Bengal.

The cats are kept caged, well fed and the atmosphere maintained at a high temperature. It is said that they produce most civet when they have been angered and the pouch beneath the animal is then scraped. The secretion appears as a soft fatty mass with a particularly disgusting odour, but when diluted it is very attractive.

4. CASTOR

This is a secretion found in preputial follicles of both the male and female beaver. When dried it is dark and has the smell of birch tar.

The value of animal perfumes is difficult to describe in scientific terms. They act as 'fixatives' preventing the rapid dissipation of lighter vegetable scents; they diffuse, giving a perfume 'life'; they impart a certain tangibility, giving it 'body'. Perhaps it would be clearer to say that it is the odours of animal origin that are dominant in the heavy French 'evening' perfumes while the flower distillates are preponderant in the lighter *parfums de jeune-fille*.

The exact chemical manner in which the animal perfumes enhance others has eluded chemists, but the physical vehicles for all these animal secretions are based upon some kind of buttery albumen-lanolin complex rich in cholesterol. This material produces many cross-linkages which may explain its remarkable fixative quality.

Natural vegetable perfumes
Generally speaking these are essential oils, but occasionally a glucoside may be responsible for the smell when it is decomposed by an enzyme as in the case of jasmine.

The perfume may be present in every part of the plant, of which the following (taken from Mauret) are examples:

Flowers: carnation, jasmine, rose, orange-blossom.
Flowers and Leaves: lavender, peppermint, violet.
Leaves and Stems: cinnamon, verbena, geranium.
Bark: canella.
Wood: cedar, santal.
Rhizomes: ginger, orris, calamus.
Roots: angelica, sassafras.
Fruit (skin): bergamot, lemon, bitter orange.
Seeds: bitter almonds, lemon, nutmeg.
Gums: labdanum, myrrh, olibanum, balsam of Peru.

The fragrance of the plant varies with different varieties of the species; with the soil, climate and method cultivation and with the atmospheric conditions existing at the time of collection. The essential oils are constituted from alcohols, esters, hydrocarbons, acids, aldehydes and ketones which originate mostly in the chloroplasts and these are formed only in the young organs of the plant, passing by diffusion from the leaf to the stalk and thence to the flower. A large proportion of these substances is used up in fecundation, possibly converted to oxygenated products with the liberation of energy, so that the flowers should be gathered before fertilization.

The odorous substances then have to be extracted from the vegetable material, and there are three main processes by which this is done: distillation, expression and dissolving out.

1. DISTILLATION
There is evidence that this method was known in 1000 B.C., but Aricenna is usually credited with the invention of the first alembic

or still. There are two ways of collecting the distillate of plants, the first is to boil them in water until it has all come down into the receiver, the second is extracting 'dry' by means of steam. The scents that lend themselves to this type of preparation are lavender, rose and orange blossom.

2. EXPRESSION

Numerous methods have always been in use such as the sponge and 'ecuelle' techniques, but today much is carried out by machine. Fruits such as lemon, lime and bitter-orange, are often treated in this manner.

3. DISSOLVING OUT

This is probably the most important technique and it is done in three ways. These are enfleurage, maceration and the use of volatile solvents.

(i) Enfleurage. The solvents here are non-volatile, usually mutton fat or lard, but vegetable oils can also be used and the process takes place at room temperatures on large glass plates. The grease absorbs the essential oils from the flowers into a perfumed pomade and this is then purified by adding an alum solution and digested with benzoin. Jasmine, orange-blossom and lily perfumes can be extracted in this way.

(ii) Maceration is a very old technique. The flowers are immersed in liquid fats or vegetable oils at a temperature somewhere between 60 to 70°C. This will extract the perfume into an oily solution. In ancient times the perfumed oils would then be ready for use, but today purification and digestion takes place as in enfleurage. Rose, violet and acacia are some of the perfumes that may be extracted in this way.

(iii) Volatile solvents. The above two methods use non-volatile solvents, either at room temperature as in enfleurage or at a higher temperature in maceration. With the advent of modern chemistry a further method has become available.

Volatile solvents such as petroleum or ether are passed slowly through flowers into a still and then round again. The process is repeated until a solid perfumed wax is obtained. This is known as the *parfum naturel solide*. It is shaken in alcohol solution for 24 hours in what is known as a batteuse and any remaining soluble wax is then separated by freezing to −20°C. Salt is added and what is called the 'Absolute Flower Oil' rises to the surface.

Often this oil is coloured and may stain, so it is discoloured by ultra-violet light or by further distillation with ethylene glycol.

It is interesting to note that 1 ton of hand-picked jasmine extracted by enfleurage will produce 1 lb of absolute oil and that 100,000,000 rose petals are required to make 1 lb of rose attar.

There are other techniques of extraction such as absorption by silica gel or carbon or butane. The butane extraction method produces very high quality absolutes known in the trade as Butaflors and is particularly useful for delicate flowers such as lilac and muguet. Extraction may be carried out at high pressures and low temperatures.

Synthetics

As could be expected, synthetic substances are now playing an increasingly important role in the manufacture of perfumes.

The first synthetic was nitro-benzene called by Mitscherlich 'essence of mirbane'. Methyl alicytate came next with its smell of oil of wintergreen. In 1868 coumarin, the perfume of new-mown hay was prepared and in 1875 the vanilla flavour vanillin. In 1898 Tiemann and Kuger produced ionone which has the odour of violets, and the odour of musk was artificially reproduced in 1888 by Baur using a derivative of s-trinitro-benzene. Since then synthetic musks and indeed virtually every other type of smell have been produced.

Attempts at producing good chemical 'fixators' to replace the costly animal perfumes are constantly being made and if they are not always as good, mixtures reinforced by resin gums or balsams are often very usable.

The technique of the perfumer's art, with its array of smells, is most difficult. By comparison painting is much easier and if compared to music the perfumer represents the composer, who must have an intimate knowledge of at least 1000 aromatic substances. He must know their odour value, source or origin, cost, the characteristics which produce their quality and their capacity to blend. He must be able to imagine the perfume in his mind, and be prepared to execute a commissioned 'piece' such as a virile, male scent for an after-shave lotion or an angelic, elusive feminine creation for a fashion house.

One of the main difficulties has been that a classification based on sources of origin is of no help as it does not represent any of the frankly emotional qualities of smells. Many attempts have been made to produce a useful, work-bench, classification. Primmel recognized 18 distinct types:

> bitter almond; ambergris; aniseed; vanilla; camphor; clove; lemon; pear; jasmine; lavender; peppermint; musk; orange-flower; rose; sandalwood; cinnamon; tuberose; violet.

Pierse compared odours with sounds. He tried to formulate an octave of odours with certain smells coinciding. Thus almond, heliotrope, vanilla and clematis blend together, each producing different degrees of a nearly similar impression, while citron, lemon, orange-peel and verbena form a higher octave.

The half-notes are represented by semi-odours such as the rose.

The perfumer must take odours together as a chord so that the perfume may be harmonious.

In 1927 Crocker and Henderson suggested that odours were made up from 4 elements of sensation:

1. Fragrant or sweet.
2. Acid or sour.
3. Burnt or empyreumatic.
4. Caprylic or enanthic.

They had a coefficient for each component and noted 8 degrees for each more or less arbitrarily, smelling for one component at a time giving numbers. Thus jasmine was 6 to 3 3 to 4 3 to 4.

Poucher initiated a trade technique based on an extraordinarily laborious estimation of duration of evaporation by smelling at 16°C. The substances whose smell lasted longest, such as patchouli and oakmoss were given the number 100. A strip was dipped in liquids up to one inch, while solids were made into 10 per cent solutions in diethyl phthalate and again smelt on a strip. Those which evaporated in less than a day were given the number 1. He then divided the result of four years' smelling never more than four a day into three groups:

> 1 to 14 Top notes—Limes, peppermint, lavender.
> 15 to 60 Middle notes—Verbena, orange-flower, jasmine.
> 61 to 100 Basic notes—Oakmoss, patchouli, musk.

The last group acting as fixers. This classification is apparently useful to the working perfumer.

FLAVOURS

Perception of flavour requires perception of the odour of foods. In the absence of the sense of smell it is only possible to experience the quality of a substance in terms of its simplest modalities of saltiness, sweetness, sourness and bitterness as well as its texture and temperature. The whole range of subtle qualities which are referred to as 'flavour' is lost when the sense of smell is absent.

It is interesting in that context to compare the patients who suffer from anosmia with those who have had both chordae tympani nerves damaged. The anosmic patient will complain mainly of loss of taste—substances like tea will be tasteless although he will be aware of its sweetness or otherwise; no pleasure can be obtained from food, and wines would be totally devoid of bouquet. On the other hand, there are numerous patients who have had damage done to the chorda tympani nerve which transmits taste sensation from the front of the tongue. This is either following ear operations or resulting from chronic middle ear infection, as the nerve passes right across the middle ear in order to join the lingual nerve to the tongue. The author tested the sense of taste of the anterior part of the tongue using salt, sugar and vinegar as well as measuring the threshold for electrical stimulation with the electrogustometer in many patients with perforated tympanic membranes. Almost every patient with sizeable damage showed some degree of taste impairment; many could not taste these substances at all. None of these patients, however, indicated the slightest concern over their enjoyment of food. Even those who had undergone acute section of the chorda in stapedectomy complained of an unpleasant feeling or metalic sensation in the tongue rather than of loss of taste.

Attempts at changing the flavour of foods are extremely old. Possibly one of the oldest is the cooking of meat. Not long after came the baking of bread and roasting of green coffee. The next stage came in attempts at preventing the loss of desirable flavours as a result of processing and storage of foods. This grew in importance as canning of foodstuff became widespread. At the present time flavour chemistry has been attempting to identify the volatile components of foods which are responsible for their odour and where this has been possible synthetic reduplication can be achieved.

Although great progress has been made in techniques of chemical identification, some of the main obstacles remain psychological and physiological. Firstly there is the tremendous difficulty of describing the delights of food. It is difficult enough to convey the beauty of a sculptured Venus without reproducing it by photographing or copying it, despite the fact that we can indicate its size and volume in absolute measures. Its colour, its temperature even and its surface texture can also be described in terms which are immediately and exactly comprehensible. If we are describing a flavour it is immediately apparent that there is no coherent nomenclature to describe these experiences. Secondly there is no comprehensive flavour theory which can be used to establish criteria for objective flavour evalutions. Attempts should nevertheless be made to consider flavours along those lines.

Evaluation of flavour

1. DESCRIPTION

If there are no specific terms in which we can describe a flavour it is necessary instead to use a collection of terms usually applied to other senses.

1. Taste

It is generally assumed that there are four modes of taste which can be perceived by the tongue. Namely salt, sweet, sour and bitter. Although it is not certain that there are not more, most substances appear to fall into these groups. A flavour may be described in these terms, and they can also be measured by determining quantitatively the chemical components contributing to these specific properties.

2. Odour

The substance can be sniffed or sipped as in wine tasting and an attempt made to describe it in terms of commonly known odours.

An objective method of recording odorous substances is to analyse the air which contains them by gas-chromatography.

3. Sight

The appearance, bulk and colour of a substance cannot be omitted from the description. A foodstuff may look so revolting that the sensation induced cannot be separated from other aspects of flavour. On the other hand, it is well known that colouring matter is often added to foods in order to enhance the sensory system's capacity to perceive and enjoy not only the visual improvement, but also taste and smell. An important section of gastronomy consists of the arrangement and presentation of foods in a way that will please the eye.

4. Kinaesthetic

This means the feel or texture of a food and the effort required from the tongue and mastication muscles to process it. These aspects of foods are often expressed as its hardness, grittiness, fibrousness, etc.

5. Auditory

Crispness or sizzling also evokes a feel of the substance which may be sometimes totally described as producing a 'snap crackle and pop'.

2. OBJECTIVE ASSESSMENT

The importance of some sort of objective record flavour becomes very great when dealing with industrial processes. Foods have to reproduce a guaranteed flavour, and various techniques have been developed to record its various qualities. None of these alone or all of them together provides enough information to assure complete reproducibility, but the following methods are commonly used:

1. Testing panel

Various panels with a number of persons on them have been used to assess flavours according to prearranged criteria. These may be descriptive, using certain reference points, or they may involve preference, if consumer acceptance is at stake. If enough substances and tasters are used, statistical procedures may be applied.

2. Gas-chromatography

The pattern of odorous components in a substance can be demonstrated by gas-chromatography. Such an investigation on cheddar cheese produced 130 components.

3. Mass spectrometry and nuclear magnetic resonance

These techniques can then be used to identify some, at least, of the components.

SMELL POLLUTION AND ITS CONTROL

The attitude of Western civilization towards noxious odours appears to have passed through 3 phases. Originally bad smells were considered fatalistically, the poor and uneducated were unwashed and smelt bad; their habitations were dirty and foul. Those who knew better took care to avoid the more insalubrious districts and kept their distance from the 'great unwashed'. This was the extent to which respect for cleanliness could take them. In the second period, which took place in the later part of the era of industrialization, an attempt was made to reconquer areas which stank of dilapidation and filth, 'cleaning-up' efforts were becoming, if not common, at least the accepted aim of right-thinking people. At the present time we are entering a new phase of realization that we are ourselves polluting our lives while, at the same time, our technology, is capable of preventing this and reversing the process.

It is obvious that in each historical period the attitude towards pollution of this type followed, rather than preceded social change. In earlier times a campaign against smell was a campaign against poverty. Now we see how the outcry against pollution has followed that against noise and how it has often copied the techniques for social agitation developed by the noise-abaters.

As legislation and therefore litigation becomes more common it will be necessary to know some of the means available for minimizing pollution. Doctors and lawyers will need to distinguish the toxic effects on the nose or body in general from the simple psychological ones of living in an environment with an unacceptable odour. This problem has already had to be faced in disputes over sound trauma. It is necessary to separate the cases of objective damage to the cochlea resulting in a special type of deafness, from the psychological trauma of repeated sonic bangs. This is not to suggest that mental anguish deserves less attention than purely physical damage. On the contrary, in many circumstances it may be the more important element, but it is a different one and must be kept separate if we are to understand and cope with the problem.

The following is a brief account of the methods available for preventing pollution by unacceptable smells.

The simplest approach has been the siting of factories, airports, etc., away from inhabited areas. It has turned out to be no solution, as this type of development has always been accompanied by housing for the workers in the industries and therefore defeated its own purpose. A second method has been simply to ban smoke-producing fires. Where it is possible there is, of course, nothing more successful, but it is only in rare cases that different non-polluting means of production are possible.

There are two forms in which odorous pollutants leave their sources:

1. Liquid waste

The effluent may be particularly unpleasant if it arises from protein food plants; fish canneries; soy, corn and wheat product factories; slaughter houses and certain chemical manufactures. There are three problems involved:

(i) Smell.

(ii) Corrosion. The contents of the liquid waste may be so corrosive that it may damage the system designated to carry it.

(iii) Toxicity. It may destroy living organisms which come into contact with it.

The oldest method of dealing with liquid waste has been to discharge it into the nearest available sewer, river or lake. The problem is thus solved unless legal intervention takes place. This usually happens only if corrosion is threatening the sewage system, or if aquatic life and shore vegetation die

Another method is to create pits, pools, tanks and other similar structures to store the waste. This system is very limited, as it can rarely be extended and there is considerable danger of seepage.

2. Vapour waste

This is much easier to handle and dispose of by the methods which will be described.

The techniques of smell control can be broadly divided into three main groups:

1. ELIMINATION OF ODOURS BY MECHANICAL AND PHYSICAL MEANS

1. Ventilation

The air contains dust particles whose size and nature depend on the locality. They may be soot or smoke; fine clay and soil; animal and vegetable matter, including the pollens, spores and bacteria which are biologically active.

The size of these particles ranges from 0·001 μm in diameter for smokes to 150 μm for dusts. Some heavy industrial dusts may be as large as 8000 μm. According to Viessman (1964) particles which are less than 0·3 μm remain suspended in still air. Those less than 10 μm remain suspended in disturbed currents, and in very disturbed air even particles of 100 μm may be retained.

Ventilation is a dilution process which depends on the addition of clean air at a rate which keeps the odorous pollutant below a Maximum Allowable Concentration (MAC). The values for the MAC levels of most pollutants have been worked out experimentally and have been established in standard tables. In order to achieve this effect, the nature of the contaminating load must be known, the quantity at which it is produced and the rate at which it is added to the air. The temperature of the air is also of importance as is the volume required to dilute to the MAC.

According to the *Guide and Data Book* (1962) of the American Society of Heating, Refrigerating and Air Conditioning Engineers the following simple formula can be of considerable value

$$\text{Quantity of ventilation air} = \frac{\text{Rate of contaminant generation} \times 10^6}{\text{MAC—Concentration in supply air}}$$

In order to attain the best result suction should be applied closest to the source of contamination.

In residential and commercial premises there is, apart from odour control, the necessity of maintaining the oxygen and carbon dioxide content of the air.

The oxygen requirement per person varies with his activity and is about 0·89 cu ft per man hour when walking at 1 mph. During moderate activity 0·74 cu ft of CO_2 per man hour are produced.

2. Combustion

Most odorous vapours and gases are hydrocarbons or organic compounds capable of being oxidized completely in air at high temperature. The resulting carbon dioxide and water vapour is of course odourless, but even when the original substances contained sulphur or nitrogen, the resulting oxides do not smell as much.

Hein (1964) pointed out that combustion methods combine in one step the function of capture, destruction and final disposal of pollutants. The only drawback is that materials cannot be recovered and re-used. Broadly speaking there are two main methods:

(i) High temperature oxidation in a combustion chamber fired directly with a naked flame.

(ii) Lower temperature oxidation using catalysts. The catalytic surface is usually platinum alloys and/or a combination of platinum and alumina. These metals are coated as a thin film on a support material such as porcelain rods or nickel-alloy ribbon.

3. Activated charcoal

The odorants in a liquid can be adsorbed onto activated charcoal and then filtered off.

4. Air-washing

This method has two advantages, it can be used to control smells and also to recover useful products. It is probably the oldest industrial technique of this type, having been developed at the beginning of the century.

The contaminated air is drawn through a large duct containing nozzles which spray water-vapour in its path. If it is then cooled condensation takes place with deposition of the odorous molecules on the water surface.

Other liquids such as oils, salt solutions or glycerol may be used to absorb the odorants. These can also be adsorbed by attachment to particulate matter.

These events can be made to take place in large odour-adsorption towers, often tiers of shelves packed with glass fibre over which water is injected in a fine spray.

5. Sorption towers

These are very useful for smelly vapours. They can be packed with silica-gel, ion-exchange resins or activated carbon.

2. ELIMINATION OF ODOURS BY CHEMICAL MEANS
1. Oxidizing

This can be done by hydrogen peroxide for liquids and by ozone for vapours.

2. Chlorination

This is a very effective method, but the smell of chlorine may itself be unpleasant and the gas is toxic if not properly controlled.

3. Ultraviolet light

Irradiation of the air in this way is sometimes used.

4. Bactericides

Formaldehyde, cresols, mercurials, acids, alkalis and quaternary ammonium chlorides have all been used to kill the bacteria responsible for putrefaction and the odours it produces.

5. Non-specific reactants

Most chemical reactants are specific for a special class or type of odorant or contaminant, and are themselves not always free from a noxious smell.

Some do not have these disadvantages. One type, described by Hanna *et al.*, (1964) consists of a combination of potassium permanganate and activated alumina in the shape of pea-sized balls. These are used in filter beds and while the alumina adsorbs the water-vapour and odorants, this is oxidized by potassium permanganate.

3. ELIMINATION OF ODOURS BY MASKING OR MODIFICATION

1. Perfuming

The simplest method is to add a stronger and better smell to the product or environment. An example is the perfuming of soap to eliminate its slightly greasy smell.

2. Olfactory anaesthesia

The temporary inactivation of the olfactory nerve endings by mild anaesthetics such as menthol is never entirely satisfactory.

3. Cancellation or compensation

Certain odorant pairs of chemicals cancel each other's smell. Examples are skatole and coumarin; ethyl mercaptan and eucalyptol. These compounds are few, however, and the technique has a limited application.

4. Chemical combination

Quaternary ammonium chlorides have a cationic N-atom with 4 covalent carbon-nitrogen bonds and a halide union. They form complexes by chemical combination with odorant molecules. The compounds of low molecular weight tend to have an irritating effect but the higher molecular weight ones are less disturbing. Other chemicals that can be used in this manner are those highly reactive double bonds and unsaturated compounds.

In urban houses the problems to be faced are rather different from the industrial environment. An interesting study of these indoor smells was made by Kenney (1954) for the American Hotel Association.

He described four sources of unpleasant smells.

1. Mildew

This was found especially in warm, humid regions and was intensified by wall-to-wall carpeting which cannot be removed for drying after shampooing. There are a number of mildewicides available to deal with this problem.

2. Lavatories

A number of masking agents are used to cover the ammoniac smell. These can be prepared as continously evaporating perfumed material or as odourized cakes. Frequent washing of floors and fixtures with disinfectant solutions such as pine oil and phenolic germicides is very useful, but there is the objection that their own odour is not always pleasant.

3. Stale smoke

This can be resolved by adequate ventilation, by spraying with odour-masking agent and by cartridges of activated carbon.

4. Kitchen

Smells from the kitchen originate from the process of cooking or from undisposed-of garbage. There should be ventilators over the cooking units and garbage should be placed in sealed containers.

Finally, one should not end this chapter without noting that in many circumstances smells should be added to gases in order to unmask their appearance. The most common example is ordinary household gas which can only be detected by its smell.

Spealman (1954) in a detailed study of the smells produced by the various chemicals used in aviation pointed out how useful these could be in detecting leaks.

Index

References

ADRIAN, E. D. (1942) Olfactory reactions in the brain of the hedgehog. *J. Physiol.*, **100**, 459.

ADRIAN, E. D. (1950) Electrical responses in the olfactory bulb. *Brit. Med. Bull.*, **6**, 1534.

ADRIAN, E. D. (1951) Responses in the olfactory bulb and secondary pathways. *J. Physiol.*, **115**, 42.

ADRIAN, E. D. (1956) Action of mammalian olfactory organs. *J. Laryng. and Otol.*, **70**, 1.

ALLEN, W. F. (1929) Effects of olfactory and trigeminal stimulation. *Am. J. Physiol.*, **88**, 117.

ALLEN, W. F. (1941) Effects of ablating pyriform-amygdaloid areas and hippocampi on conditioned reflexes. *Am. J. Physiol.*, **132**, 81.

ALLEN, W. F. (1943) Distribution of cortical potentials from olfactory stimulation. *Am. J. Physiol.*, **139**, 553.

ALLISON, A. C. (1950) Investigation of the morphology of the mammalian olfactory system. Thesis. Oxford.

ALLISON, A. C., (1953) Structure of olfactory bulb and pathways in the rabbit and rat. *J. Comp. Neurol.*, **98**, 309.

ALLISON, A. C. and WARWICK, R. T. T. (1949) Quantitative observations on the olfactory system of the rabbit. *Brain*, **72**, 186.

ALVAREZ, W. C. (1966) Changes in the brain with advancing age. *Geriatrics*, **21**, 111.

AMOORE, J. E. (1952) Stereochemical specificities of human olfactory receptors. *Perf. Essent. Oil Record*, **43**, 321.

AMOORE, J. E. (1962) Stereochemical theory of olfaction. 1. Identification of seven primary odours. *Proc. Sci. Sect., Toilet Goods Assn., Special Supplement to No. 37*, 1.

AMOORE, J. E. (1962) Stereochemical theory of olfaction. 2. Elucidation of the Stereochemical properties of the olfactory receptor sites. *Ibid.*, 13.

AMOORE, J. E. (1963) Stereochemical theory of olfaction. *Nature*, **198**, 271.

AVERY, T. L. (1969) Pheromone-induced changes in the mouse pituitary. *Science*, **164**, 42.

BANG, B. G. (1964) The mucous glands of the developing human nose. *Acta anatomica (Basle)*, **59**, 297.

BECK, L. H. and MILES, W. R. (1947) Theoretical experimental relationships between infrared absorption and olfaction. *Science*, **106**, 511.

BEDICHEK, R. (1960) *The Sense of Smell*. London: Michael Joseph.

BEIDLER, L. M. (1957) Facts and theory on taste and odor perception in chemistry of natural food flavours. J. H. Mitchell, Jr. Ed. 7-43.

BEIDLER, L. M. and TUCKER, D. (1955) Response of nasal epithelium to odor stimulation. *Science*, **122**, 76.

BIEBER, I. (1959) Olfaction in sexual development. *Am. J. Psychol.*, **13**, 851.

BLOOM, G. and ENGSTRÖM, H. (1953) Interciliary structures in the epithelium of the respiratory tract. *Ann. Otol. Rhin. and Laryng.*, **62**, 15.

BOZZA, G., CALEARO, C., TEATINI, P. (1960) On the making of a rational olfactometer. *Acta Oto-Laryng. (Stockholm)*, **52**, 189.

BREUNINGER, H. (1964) Über das physikalisch chemische Verhalten des Nasenschleims. *Arch. Ohr.-, Nas.-, u. KehlkHeilk.*, **184**, 133.

BRONSON, F. H. and CAROOM, D. (1971) Preputial glands of the male mouse. *J. Reprod. Fertil.*, **25**, 279.

BROOKSBAND, B. W. L. and HAZLEWOOD, G. A. D. (1950) A musky steroid in human urine. *Biochem. J.*, **47**, 36.

BROWN, E. A. (1967) Measurement of air resistance in the nasal passages. *Rev. Allergy*, **21**, 852.

BRUCE, H. M. (1959) An exteroceptive block to pregnancy in the mouse. *Nature (London)*, **184**, 105.

BRUCE, H. M. (1970) Pheromones. *Brit. Med. Bull.*, **26**, 10.

BRUNN, A. (1892) *Arch. Mikr. Anat.*, **39**, 632.

CARR, W. J., LOEB, L. S., DISSINGER, M. L. (1965) Responses of rats to sex odours. *J. Comp. Physiol. Pyschol.*, **59**, 370.

CARRERAS, M., MANCIA, D., MANCIA, M. (1967) Centrifugal control of the olfactory bulb. *Electroenceph. Clin. Neurophysiol.*, **23**, 190.

CHAVANNE, L. (1937) Sécrétion nasale et glandes endocrines. *Ann. Oto-Laryng. (Paris)*, **5**, 401.

CHOUARD, C. (1967) La physiologie de l'olfaction. *Cah. Coll. Méd. Hôp. Paris*, **8**, 981.

CLARK, LE GROS (1951) Olfactory pathways. *Journ. Neurol. Neurosurg. Psychiat.*, **14**, 1.

CLARK, LE GROS (1956) Structure and organization of olfactory receptors in the rabbit. *Yale J. Biol. and Med.*, **29**, 83.

CLARK, LE GROS and WARWICK, R. T. T. (1946) Olfactory pathways. *Journ. Neurol. Neurosurg. Psychiat.*, **9**, 101.

CLUZEL, J. (1964) Contribution de l'olfactométrie aux syndromes endocriniens. Thèse. Marseille.

CROCKER, E. C. and HENDERSON, L. F. (1927) Analysis and classification of odours. *Am. Perfumer and Essent. Oil Rev.*, **22**, 325.

CROSS, B. A. and HARRIS, G. W. (1952) Milk ejection reflex. *J. of Endocrinol.*, **8**, 148.

DALHAMN, T. and RHODIN, J. (1956) Mucous flow and ciliary activity in exposure to gas. *Brit. J. Indust. Med.*, **13**, 110.

DAVIES, J. T. (1962) The Mechanism of olfaction. *Symp. Soc. Expt. Biol.*, **16**, 170.

DAVIES, J. T. and TAYLOR, F. (1959) Adsorption in olfaction. *Biol. Bull.*, **117**, 222.

DAWES, J. D. K. and PRICHARD, M. M. L. (1953) Vascular arrangements of nose. *J. Anat.*, **87**, 311.

DE LORENZO, A. (1970) Olfactory neurone. In *Taste and Smell*. Ed. Wolstenholme, G. and Knight, J., London: Churchill.

DEMORSIER, G. (1962) La dysplasie olfactogénitale. *Acta Neuropath. (Berlin)*, **1**, 433.

DONOVAN, B. T. and KOPRIWA, P. C. (1965) Removal or stimulation of olfactory bulbs on estrous cycle of guinea pig. *Endocrinology*, **77**, 213.

DOUEK, E. E. (1967) Smell: recent theories and their clinical applications. *J. Laryng.*, **81**, 431.

DOUEK, E. E. (1907) Some abnormalities of smell. *J. Laryng.*, **84**, 1185.

DOVING, K. B. (1967) Comparative electrophysiological experiments on the olfactory tract. *J. Comp. Neurol.*, **131**, 365.

DURRANS, T. H. (1920) Odour theory. *Perf. Ess. Oil Rec.*, **11**, 391.
DYSON, G. M. (1938) Scientific basis of odour. *Chem. and Ind.*, **16**, 647.

EFRON, R. (1956) Effect of olfactory stimuli on uncinate fits. *Brain (London)*, **79**, 267.
EFRON, R., (1957) Conditioned inhibition of uncinate fits. *Brain (London)*, **80**, 521.
ELLIS, HAVELOCK (1928) *Studies in the Psychology of Sex.* Philadelphia: F. H. Davis Company.
ELSBERG, C. A., LEVY, I., BREWER, E. D. (1936) New method for testing smell. *Science*, **83**, 211.
ELSBERG, C. A. and STEWART,   (1938) Quantitative olfactory tests in the diagnosis of brain tumours. *Arch. Neurol. and Psychiat.*, **40**, 471.
ENGEN, T. (1961) Direct scaling of odor intensity. *Rept. Psychol. Lab. Univ. Stockholm*, No. 106.
EWERT, G. (1965) Mucus flow rate in the human nose. *Acta Otolaryng. (Stockholm)*, Suppl. **200**, 1.
EYFERTH, K. and BALTES, P. B. (1969) Über Normierungseffekte in einer Faktoreanalyse. *Z. Exp. Angew. Psychol.*, **16**, 38.

FALCONER, M. A. and TAYLOR, D. C. (1968) Treatment of Epilepsy. *Arch. Neurol. (Chicago)*, **19**, 353.
FALCONER, M. A. and TAYLOR, D. C. (1970) Surgery in temporal lobe epilepsy. *J. Neurosurg.*, **33**, 233.
FALCONER, M. A., SERAFETINIDES, E. A., CORSELLIS, J. A. (1964) Temporal lobe epilepsy. *Arch. Neurol. (Chicago)*, **10**, 233.
FAWCETT, D. W. and PORTER, K. R. (1954) Structure of ciliated epithelia. *J. Morphol.*, **94**, 221.
FRANCK, H. (1966) Répercussions de l'ablation des bulbes olfactifs. Thèse. Strasburg.
FRISCH, D. (1967) Ultrastructure of mouse olfactory mucosa. *Amer. J. Anat.*, **121**, 87.

GARTEN, S. (1903) *Beiträge zur Physiologie der marklosen Nerven.* Jena, Germany: Gustav Fischer.
GASSER, H. S. (1956) Olfactory nerve fibres. *J. Gen. Physiol.*, **39**, 473.
GERRARD, R. W. and YOUNG, J. L. (1937) Electrical activity in central nervous system of frog. *Proc. Roy. Soc. (B)*, **122**, 343.
GESTELAND, R. C. (1965) Chemical transmission in the nose of the frog. *J. Physiol.*, **181**, 525.
GOLDING-WOOD, P. H., (1961) The vidian nerve. *J. Laryng.*, **75**, 232.
GOWERS, W. R. (1881) Epilepsy and other Chronic Convulsive Diseases. London.
GRAY, J. (1928) *Ciliary Movement.* Cambridge University Press.
GUERRIER, Y., RLU, R., LEONARDELLI, G., LE DEN, R., PIZZETTI, F. (1969) Tumeurs du neuroépithélium olfactif, p. 91. In *Les Tumeurs Nerveuses en ORL.* Montpellier: Monographie Chauvin-Blache.

HENNING, H. (1924) *Smell.* 2nd Ed. Leipzig:
HEYNIX, M. (1933) Physiologie de l'olfaction. *Rev. Oto-neuro-ophtalm.*, **11**, 10.
HUGHES, J. R. and HENDRIX, D. E. (1964) Olfactory coding. *Nature*, **216**, 404.

REFERENCES

IOANESCU, V. (1961) Olfactory and vestibular sensory impulses and temporal lobe epilepsy. *Acta Psychiat. Neurol. Scand.*, **36**, 415.

JACKSON, J. H. and BEEVOR, C. E. (1888) Tumor of the temporosphenoidal lobe and the localisation of the sense of smell. *Brain (London)*, **12**, 346.
JACKSON, R. T. (1960) Olfactory pigment. *J. Cell. Comp. Physiol.* **55**, 143.
JOHNSTON, J. B., (1923) Evolution of the fore-brain. *J. Comp. Neurol.*, **35**, 337.
JOHNSTON, J. W. (1963) Application of steric theory. *Georgetown Med. Bull.*, **17**, 40.

KITTEL, G. (1968) Die moderne Olfactometrie und Odorimetrie. *Z. Laryng. Rhinol.*, **47**, 893.
KOHNE, G. (1947) Die Beziehungen des angeborenen Olfactorius Defekts zum Eunuchoidismus. *Virchows Arch. Path. Anat. (Berlin)*, **314**, 345.
KRISCH, W. (1933) Eine neue Theorie der Sinnesorgane. *Naturwiss.*, **21**, 876.

LANDAU, E. (1942) La voie sensitive olfactive périphérique. *Confinia neurol. (Basle)*, **5**, 225.
LARSELL, O. (1918) Studies in the nerve terminals. *J. Comp. Neurol.*, **30**, 3.
LE MAGNEN, J. (1953) L'olfaction et les régulations psycho-physiologiques. *J. Physiol. (Paris)*, **45**, 285.
LE MAGNEN, J. (1961) Les Fonctions Proprement Olfactives du Rhinencéphale in *Physiologie et Pathologie du Rhinencéphale*. Paris: Manon.
LE MAGNEN, J. (1965) L'Analyse Discriminative de l'Appareil Olfactif. *Rev. Laryng. (Bordeaux)*, Suppl. **86**, 827.
LEWIN, W. (1966) *Management of Head Injuries*. London: Baillière, Tindal and Cassel.
LOHMAN, A. H. M. and LAMMERS, H. J. (1963) Connections of the olfactory bulb. *Progr. Brain Res.*, **3**, 149.
LUCAS, A. M. and DOUGLAS, L. C. (1934).
LUCRETIUS, TITUS CARUS (95–55 BC) *Lucretius on the Nature of the Universe*. Harmondsworth: Penguin Books.

MACCORD, C. P., and WITHERIDGE W. N. (1949) Odors. New York: McGraw Hill.
MACKENZIE, D. (1923) *Aromatics and the Soul*. London: Heinemann.
MACLEAN, P. D. (1952) Physiologic Studies. EEG. *Clin. Neurophysiol.*, **4**, 407.
MARGERISON, J. H. and CORSELLIS, J. A. (1966) Epilepsy and the temporal lobes. *Brain*, **89**, 499.
MARR, J. N. and GARDNER, L. E. (1965) Early olfactory experience and later behaviour. *J. Genet. Psychol.*, **107**, 167.
MÉLON, J. and SCHOFFENIELS, E. (1966) Modification des sécrétions nasales. *Acta Alleng. (Copenhagen)* **21**, 497.
MICHAEL, R. P. (1965) Endocrine control of sexual activity in primates. *Proc. Roy. Soc. Med.*, **58**, 595.
MICHAEL, R. P. and KEVERNE, E. B. (1968) Pheromones in the communication of sexual status in primates. *Nature (London)*, **218**, 746.
MICHAEL, R. P. and KEVERNE, E. B. (1970) Primate sex pheromones of vaginal origin. *Nature (London)*, **225**, 84.

MICHAEL, R. P. and SAAYMAN, G. (1967) Sexual performance index of male rhesus monkeys. *Nature (London)*, **214**, 425.

MONCRIEF, R. W. (1944) *Manufacturing Chemist*, **15**, 443.

MONCRIEF, R. W. (1949) Amer. Perfum., 453.

MOULTON, D. (1960) Studies in Olfactory Acuity. *Animal Behav.*, **8**, 129.

MOULTON, D. (1962) Pigment and olfactory mechanism. *Nature (London)*, **195**, 1312.

MOULTON, D. G. and BIDLER, L. M. (1967) Structure and function of the peripheral olfactory system. *Physiol. Rev.*, **47**, 1.

MOZELL, M. M. (1964) Sorption as a mechanism of olfactory analysis of vapours. *Nature (London)*, **203**, 1181.

NEGUS, V. E. (1958) *Comparative Anatomy of Nose and Sinuses*. Edinburgh: Livingstone.

OTTOLENGHI, S. (1888) L'Olfato nei criminali. *Giorn. Med. Accad. Torino*, **36**, 427.

OTTOSON, D. (1956) Analysis of electrical activity of olfactory epithelium. *Acta Physiol. Scandinav. (Stockholm)* Suppl. 122, **35**, 1.

OTTOSON, D. (1959) Potentials in the rabbit's olfactory bulb and mucosa. *Acta Physiol. Scandinav. (Stockholm)*, **47**, 136.

PASSOUANT, P. and PTERNITIS, C. (1965) Activation de l'hippocampe. *Acta Physiol. Acad. Sci. Hung.*, **26**, 123.

PROETZ, A. W. (1924) Exact olfactometry. *Ann. Otol. Rhinol. and Laryngol.*, **33**, 275.

PROETZ, A. W. (1953) Respiratory air currents. *J. Laryng. and Otol.*, **67**, 1.

PRYSE PHILLIPS, M. (1968) Thesis, Birmingham Univ.

REES, L. (1964) Factors in vasomotor rhinitis. *J. Psychosom. Psychosom. Res.*, **8**, 101.

RESSLER, R. H., CIALDINI, R. B., GHOCA, M. L. (1968). Alarm pheromone in the earthworm. *Science*, **161**, 597.

RIMMEL, E. (1865) *The Book of Perfumes*. London: Chapman and Hall.

ROPARTZ, P. (1966) Secretio odorante sudoripares chez la souris. *C.R. Acad. Sci. (D) (Paris)*, **263**, 525.

ROPARTZ, P. (1968) The relation between olfactory stimulation and aggressive behaviour in mice. *Anim. Behaviour*, **16**, 97.

ROSEBURG, B. (1968) Die Möglichkeiten der Olfactometrie aus klinischer Sicht. *HNO (Berlin)*, **16**, 302.

ROWBOTHAM, G. F. (1966) Acute Injuries of the Head. Edinburgh: Churchill Livingstone.

SANTORELLI, G. and MAROTTA, (1965) Olfaction in epilepsy. *Rev. Laryng. (Bordeaux)*, **86**, 954.

SCHULTZE, M. (1856) Über die Endigungsweise der Geruchsnerven. *Mber. Akad. Wiss. (Berlin)*, **504**.

SEMERIA, C. (1956) Olfattometria obbiettiva. *Minerva otorinolaring.*, **6**, 97.

SEM-JACOBSEN, C. W., BICKFORD, R. G., DODGE, H., PETERSEN, M. C. (1953)

Human olfactory responses recorded by depth electro-encephalography. *Proc. Staff. Meeting Mayo Clin.*, **28**, 166–170.

SIGNORET, J. P. and MAULEON, P. (1961) Ablation des bulbes chez la truie. *Ann. Biol. Animal Biochem. Biophys.*, **2**, 67.

SKRAMLICK, E. VON (1936) *Handbuch der Physiologie der Niederen Sinne.* Leipzig.

SPILLANE, J. (1938) Olfactory alloaesthesia. *Brain*, **61**, 393.

STEVENS, S. S. (1951) *Handbook of Experimental Psychology*, New York: Wiley.

STEVENS, S. S. (1961) Psychophysics of Sensory Function in *Sensory Communication*. W. A. Rosenblith (ed.), 1: 33. Massachusetts Institute of Technology Press and J. Wiley, New York.

STOKSTED, P. (1952) Cycle of the nose. *Acta oto-laryng.* (*Stockholm*), **42**, 175.

STOKSTED, P. (1953) Resistance in the nose. *Acta oto-laryng.* (*Stockholm*), Suppl. 109.

STUIVER, M. (1958) *Biophysics of Smell.* Thesis, Gröningen.

STUIVER, M. (1960) An olfactometer. *Acta Otolaryng.*, **5**, 135.

SUMNER, D. (1962) Testing smell. *Lancet*, ii, 895.

SUMNER, D. (1964) Post-traumatic anosmia. *Brain*, **87**, 107.

TAKAFI, S. F., WYSE, G. A., YAJIMA, T. (1966) Permeability of olfactory receptive membrane. *J. Gen. Physiol.*, **50**, 473.

TEUDT, H. (1920) *Chem. Abstracts*, **14**, 1685.

THIELLEMENT, R. (1955) *Rapports Génitaux Olfactifs.* Thesis. Maisons-Alfort, No. 30.

THIESSEN, D. D., FRIEND, H. C., LINDZEY, G. (1968) Androgen Control of Territorial Marking in the Mongolian Gerbil. *Science*, **160**, 432.

THOMSON, A. and DUDLEY BUXTON, L. H. (1923) *J. Roy. Anthrop. Inst.*, **53**, 92.

TOULOSE, E. and VASCHIDE, N. (1899) Mesure de l'odorant chez l'homme et la femme. *C.R. Soc. Biol.*, **51**, 381.

TUCKER, D. and SHIBUYA, T. (1965) Olfactory receptors. *Symp. Quant. Biol.*, **30**, 207.

UNGERER, W. G. and STODDARD, R. B. (1922) Odor value analysis. *Ungerer's Bull.*, 3 No. **1**, 7.

VALENTA, J. G. and RIGBY, M. K. (1968) Discrimination of the odor of stressed rats. *Science*, **161**, 599.

VASCHIDE, N. (1901) L'Expérience de webor et l'olfaction en milieu liquide. *C.R.* **53**, 165.

VIESSMAN, W. (1964) Ventilation control of odor. *Ann. New York Acad. Sci.*, **116**, 603.

WENZEL, B. M. (1948) Techniques in olfactometry. *Physiological Bulletin*, **45**, 231.

WENZEL, B. M. and SIEK, N. H. (1966) Olfaction. *Ann. Rev. Physiol.*, **28**, 381.

WHITTEN, W. K. (1956) Modification of oestrus cycle in the mouse by external stimuli. *J. Endocrinol.*, **13**, 399.

WIENER, J. S. (1954) *Am. J. Physical Anthropol.*, **12**, 1.

WILLEMOT, J. (1971) La muqueuse olfactive chez l'homme. *Acta Otolaryngol.* (*Stockholm*), **7**, 197.

WOODROW, H. and KARPMAN, B. (1917) A new olfactometric technique. *J. Exp. Psychol.*, **2**, 431.

WRIGHT, R. D. (1966) Why is an odour? *Nature (London)*, **209**, 551.

WRIGHT, R. D. (1966) Odour and molecular vibration. *Nature (London)*, **209**, 571.

WRIGHT, R. H. and ROBSON, A. (1969) Basis of odour specificity. *Nature (London)*, **222**, 290.

ZONDEK, B. (1967) Effects of External Stimuli on Reproduction. *CIBA Symp.* Ed. Wolstenholme and O'Connor.

ZWAARDEMAKER, H. (1904) Präzisions-Olfactometrie. *Arch. Laryng. Rhinol.*, **15**, 171.